Lecture Notes in Mathematics

Edited by A. Dold and B. Eckmann

Series: Mathematisches Institut der Universität Erlangen-Nürnberg
Advisers: H. Bauer and K. Jacobs

823

Josef Král

Integral Operators in Potential Theory

Springer-Verlag
Berlin Heidelberg New York 1980

Author

Josef Král
Matematický ústav
Žitná 25
11567 Praha 1
ČSSR

AMS Subject Classifications (1980): 31 B10, 31 B20, 35 J05, 35 J25, 45 B05, 45 P05

ISBN 3-540-10227-2 Springer-Verlag Berlin Heidelberg New York
ISBN 0-387-10227-2 Springer-Verlag New York Heidelberg Berlin

Library of Congress Cataloging in Publication Data. Král, Josef, DrSc. Integral operators in potential theory. (Lecture notes in mathematics; 823) Bibliography: p. Includes indexes. 1. Potential, Theory of. 2. Integral operators. I. Title. II. Series: Lecture notes in mathematics (Berlin); 823.
QA3.L28. no. 823. [QA404.7]. 510s. [515.7] 80-23501

Printing and binding: Beltz Offsetdruck, Hemsbach/Bergstr.
2141/3140-543210

CONTENTS

Introductory remark

We shall be concerned with relations of analytic properties of classical potential theoretic operators to the geometry of the corresponding domain in the Euclidean m-space R^m, $m \geq 2$.

Let us recall that a function h is termed harmonic in an open set $G \subset R^m$ if it is twice continuously differentiable in G and satisfies there the so-called Laplace equation

$$\Delta h = \sum_{i=1}^{m} \partial_i^2 h = 0 ,$$

where ∂_i denotes the partial derivative with respect to the i-th variable. (In fact, such a function h is necessarily infinitely differentiable and even real-analytic; this is usually proved in elementary theory of harmonic functions on account of the Poisson integral which will be derived in the example following theorem 2.19.) If we try to determine a harmonic function in $R^m \setminus \{0\}$ of the form $h(x) = \omega(|x|)$, where ω is an unknown function with a continuous second derivative in $]0,+\infty[\subset R^1$, we obtain an ordinary differential equation

$$\frac{d^2\omega(r)}{dr^2} + \frac{m-1}{r}\frac{d\omega(r)}{dr} = 0$$

whose solutions are

$$\omega(r) = \begin{cases} \alpha\, r^{2-m} + \beta & \text{in case } m > 2 , \\ \alpha \log r + \beta & \text{in case } m = 2 , \end{cases}$$

where α, β are arbitrary constants.

Let us denote by $A \equiv A_m$ the area of the unit sphere in R^m, i.e.

$$A_m = \frac{2\,\pi^{\frac{1}{2}m}}{\Gamma(\frac{1}{2}m)}$$

where $\Gamma(.)$ is the Euler gamma-function, and put for $x \in$ $\in R^m \setminus \{0\}$

$$h_o(x) = \begin{cases} \frac{1}{(m-2)A}\, |x|^{2-m} & \text{if } m > 2 , \\ \frac{1}{A} \log \frac{1}{|x|} & \text{if } m = 2 . \end{cases}$$

As we shall observe later, with this normalization - h_o represents a fundamental solution of the Laplace equation which means that, in the sense of distribution theory,

$$-\Delta h_o = \delta_o ,$$

where δ_o is the Dirac measure (= unit point mass situated at the origin). For $m = 3$ the function h_o occurred first in physics; according to the Newton gravitation law (or Coulomb's law in electrostatics) the vector-valued function

$$x \mapsto \text{grad } h_o(x)$$

describes the force-field of a point mass (or a point charge) placed at the origin.

The function h_z defined by

$$h_z(x) = \begin{cases} h_o(x-z) & \text{for } x \neq z , \\ +\infty & \text{for } x = z , \end{cases}$$

is sometimes called the fundamental harmonic function with pole at z. It follows from the above remark that h_z is, up to additive and multiplicative constants, the only harmonic function in $R^m \setminus \{z\}$ whose values depend on the distance from z only. If $G \subset R^m$ is open, then all the functions h_z with $z \in R^m \setminus G$ as well as their directional derivatives

$$x \longmapsto n \cdot \text{grad}_x \, h_z(x)$$

(where $n \in R^m$) are harmonic in G. The idea of using "combinations" of these simple functions for generating more complicated harmonic functions in G is classical. By "combinations" here we mean not only discrete combinations but in general the integrals of the form

(1) $$\mathcal{U}\nu(x) = \int_{R^m} h_z(x) d\nu(z) ,$$

(2) $$W\nu(x) = \int_{R^m} n(z) \cdot \text{grad}_x \, h_z(x) d\nu(z) ,$$

where ν is a signed measure with support in $R^m \setminus G$ and $z \longmapsto n(z)$ is a suitable vector-valued function with values in R^m. In classical potential theory G is usually supposed to have a smooth boundary B with area element ds and ν is taken in the form $d\nu = f ds$, where f is an appropriate function on B , while $n(z)$ is usually chosen as the unit normal to B . The integrals of the form (2) are then called the double layer potentials and proved to be useful in connection with the Dirichlet problem which reads as follows: Given a continuous function g on B , determine a harmonic function h in G such that $\lim_{\substack{x \to z \\ x \in G}} h(x) = g(z)$ for every $z \in B$. If h is taken in the form of a double layer potential (2) with the above described specification for n and $d\nu = f ds$, then evaluation of the limit at $z \in B$ leads to an integral equation of the second kind

$$f(z) + \int_B K(z,y) f(y) ds(y) = 2g(z)$$

for the unknown density f . In a similar way, the so-called single layer potentials (1) are useful in treating the Neumann problem which is formulated as follows: Given a function g on B , determine a harmonic function h in G such that $\lim_{\substack{x \to z \\ x \in G}} n(z) \cdot \operatorname{grad} h(x) = g(z)$ for every $z \in B$, where $n(z)$ is the unit exterior normal to G at z . If one tries $h = \mathcal{U}\nu$ with $d\nu = f ds$, this problem again reduces to an integral equation of the second kind for the unknown density f and the kernel of the corresponding integral operator is transposed to the kernel resulting from the Dirichlet problem for the complementary domain. Historically it was this method of treating boundary value problems in potential theory that led to the development of the Fredholm theory of equations of the second kind. In its classical formulation the method is tied up with certain à priori smoothness restrictions on the boundary of the domain, because the normal derivative occurs in the definition of double layer potentials and in the formulation of the Neumann problem. These restrictions may be entirely avoided, however, if the normal derivative is characterized weakly. Normal derivatives of single layer potentials as well as double layer potentials may then be introduced and investigated for general open sets $G \subset R^m$ without any à priori restrictions on the boundary. Some results in this direction together with their applications to boundary value problems will be described below.

§ 1

Weak normal derivatives of potentials

We shall denote by $\mathcal{D} \equiv \mathcal{D}(R^m)$ the class of all infinitely differentiable functions with compact support in R^m.

1.1. Definition. Let h be a harmonic function in an open set $G \subset R^m$ and suppose that

$$\int_P |\text{grad } h(x)|\, dx < \infty$$

for every bounded open set $P \subset G$. Then $Nh \equiv N^G h$ will denote the functional over $\mathcal{D}$ defined by

$$\langle \varphi, Nh \rangle = \int_G \text{grad } \varphi(x) \cdot \text{grad } h(x)\, dx\,, \qquad \varphi \in \mathcal{D}\,.$$

Nh will be termed the generalized normal derivative of h.

Remark. The reason for this terminology lies in the fact that, in the case when G is bounded by a smooth closed surface B with area element ds and exterior normal $n = (n_1, \ldots, n_m)$ and when the partial derivatives $\partial_i h$ $(i = 1, \ldots, m)$ extend from G to continuous functions on the whole $G \cup B$, the Gauss-Green formula yields

$$\langle \varphi, Nh \rangle = \int_B \varphi\Big(\sum_{i=1}^m n_i\, \partial_i h\Big) ds\,, \qquad \varphi \in \mathcal{D}\,.$$

Consequently, Nh is a natural weak characterization of the normal derivative $\sum_{i=1}^{m} n_i \, \partial_i h = \frac{\partial h}{\partial n}$.

1.2. Remark. If G and h have the meaning described in the definition 1.1, then $\langle \varphi, N^G h \rangle = 0$ for every $\varphi \in \mathcal{D}$ whose support does not meet the boundary of G. In other words, the support of $N^G h$ is contained in the boundary of G.

Proof. Suppose that the support of $\varphi \in \mathcal{D}$ does not meet the boundary of G and define $\tilde{\varphi}$ so that $\tilde{\varphi} = \varphi$ in G, $\tilde{\varphi} = 0$ on $R^m \setminus G$. Clearly, $\tilde{\varphi} \in \mathcal{D}$ and if $\tilde{h}$ is any twice continuously differentiable function on R^m coinciding with h near the support of $\tilde{\varphi}$, then

$$\int_G \operatorname{grad} \varphi(x) \cdot \operatorname{grad} h(x) dx = \int_{R^m} \operatorname{grad} \tilde{\varphi}(x) \cdot \operatorname{grad} \tilde{h}(x) dx =$$

$$= - \int_{R^m} \tilde{\varphi}(x) \Delta \tilde{h}(x) dx = 0 ,$$

because $\tilde{\varphi} \Delta \tilde{h} = 0$ everywhere.

1.3. Notation. The ball of radius r and center y in R^m will be denoted by

$$\Omega(r,y) \equiv \Omega_r(y) = \{ x \in R^m ; \ |x-y| < r \} .$$

For $M \subset R^m$ we denote by $\operatorname{diam} M$ the diameter of M, by $\operatorname{cl} M$ the closure of M, and by $\mathcal{H}_k(M)$ the (outer) k-dimensional Hausdorff measure of M. Let us recall that

$$\mathcal{H}_k(M) = \lim_{\varepsilon \to 0+} \mathcal{H}_k^{\varepsilon}(M) ,$$

where

$$\mathcal{H}_k^{\varepsilon}(M) = 2^{-k} V_k \inf \sum_n (\operatorname{diam} M_n)^k$$

with the infimum taken over all sequences of sets $M_n \subset R^m$ such that $\operatorname{diam} M_n \leqq \varepsilon$ and $\bigcup_n M_n \supset M$ and with

$$V_k = \frac{\pi^{\frac{k}{2}}}{\Gamma(\frac{k}{2}+1)}$$

equal to the volume of the unit ball in k-space. The normalization is chosen in such a way that $\mathcal{H}_m(M)$ coincides with the outer m-dimensional Lebesgue measure of M ; if M is a simple smooth k-dimensional surface in R^m , then $\mathcal{H}_k(M)$ coincides with the area of M . (Basic facts concerning Hausdorff measures may be found in the monograph [Ro] .)

By a signed measure we mean a finite σ-additive set function defined on the σ-algebra of Borel sets in R^m . If ν is a signed measure and $M \subset R^m$ is a Borel set, then $|\nu|(M)$ denotes the total variation of ν on M ; we put $\|\nu\| = |\nu|(R^m)$. If $B \subset R^m$ is compact, we denote by $\mathcal{C}'(B)$ the linear space of all signed measures ν with $|\nu|(R^m \setminus B) = 0$, i.e. with support in B ; $\mathcal{C}'(B)$ is a Banach space if equipped with the norm $\|\ldots\|$. The abbreviation spt ν (spt φ, ...) will denote the support of ν (support of φ, ...). If $\nu \in \mathcal{C}'(B)$, then the potential $\mathcal{U}\nu(x)$ defined by (1) is meaningful for all $x \in R^m \setminus B$ and represents a harmonic function in $R^m \setminus B$. The symbol δ_y will denote the Dirac measure defined by

$$\delta_y(M) = \begin{cases} 1 & \text{if } y \in M , \\ 0 & \text{if } y \notin M \end{cases}$$

on Borel sets $M \subset R^m$. Thus $\mathcal{U}\delta_y = h_y$ on R^m . We put $\Gamma = \{x \in R^m;\ |x| = 1\}$, so that $A = \mathcal{H}_{m-1}(\Gamma)$.

1.4. Remark. The following elementary transformation formula will be often useful below:

If g is an integrable function on R^m and $z \in R^m$ is fixed, then the function

$$\theta \longmapsto \int_0^\infty t^{m-1} g(z+t\,\theta)\,dt$$

is defined for $\mathcal{H}_{m-1}$-almost every $\theta \in \Gamma$, is integrable $d\mathcal{H}_{m-1}$ and

$$\int_{R^m} g(x)\,dx = \int_\Gamma \Big(\int_0^\infty t^{m-1} g(z+t\,\theta)\,dt\Big)\,d\,\mathcal{H}_{m-1}(\theta) .$$

Remark. If $B \subset R^m$ is compact and $\nu \in \mathcal{C}'(B)$, then for $x \in R^m \setminus B$

$$|\operatorname{grad} \mathcal{U}\nu(x)| \leqq \frac{1}{A} \int_B |x-z|^{1-m}\, d|\nu|(z) ,$$

whence we get for any bounded Borel set $P \subset R^m \setminus B$

$$\int_P |\operatorname{grad} \mathcal{U}\nu(x)|\,dx \leqq \frac{1}{A} \int_B \Big(\int_P |x-z|^{1-m}\,dx\Big)\,d\,|\nu|(z) \leqq$$

$\leqq \operatorname{diam}(P \cup B)\|\nu\| < \infty$.

We see that if $G \subset R^m$ is an open set with a compact boundary B, then $N^G \mathcal{U} \nu$ (taken in the sense of the definition 1.1) is available for every $\nu \in \mathcal{C}'(B)$.

<u>Example.</u> Fix $z \in R^m$ and let $G = R^m \setminus \{z\}$, $B = \{z\}$, $\nu = \delta_z$. Employing the transformation formula in 1.4 one gets easily for $\varphi \in \mathcal{D}$

$$\langle \varphi, N^G \mathcal{U} \delta_z \rangle = \int_{R^m \setminus \{z\}} \operatorname{grad} \varphi(x) \cdot \operatorname{grad} h_z(x) dx =$$

$$= \frac{1}{A} \int_{R^m} |z-x|^{1-m} \operatorname{grad} \varphi(x) \cdot \frac{z-x}{|z-x|} dx = \varphi(z) .$$

We see that $N^G \mathcal{U} \delta_z = \delta_z$ in this case.

Noting that

$$\int_{R^m} \operatorname{grad} \varphi(x) \cdot \operatorname{grad} h_z(x) dx = - \int_{R^m} \Delta \varphi(x) h_z(x) dx$$

we may rewrite the above equality in the form

$$\int_{R^m} \Delta \varphi(x) h_z(x) dx = - \varphi(z) , \quad \varphi \in \mathcal{D} .$$

This means that, in the language of distribution theory, $\Delta h_z = - \delta_z$.

<u>1.5. Observation.</u> If $G \subset R^m$ is an open set with a compact boundary B and $\nu \in \mathcal{C}'(B)$ then, for any $\varphi \in \mathcal{D}$,

$$(3) \qquad \langle \varphi, N^G \mathcal{U} \nu \rangle = \int_B \langle \varphi, N^G \mathcal{U} \delta_y \rangle d\nu(y) .$$

Proof. Fix $\varphi \in \mathcal{D}$ and put $c = \sup |\mathrm{grad}\ \varphi(x)|$, $P = G \cap \mathrm{spt}\ \varphi$. Elementary calculation yields the estimate

$$(4_1) \qquad \iint_{G \times B} |\mathrm{grad}\ \varphi(x) \cdot \mathrm{grad}\ h_y(x)| \, dx \, d|\nu|(y) \leq$$

$$\leq c \ \mathrm{diam}\ (P \cup B) \|\nu\|$$

which shows that the double integral

$$(4_2) \qquad \iint_{G \times B} \mathrm{grad}\ \varphi(x)\ \mathrm{grad}\ h_y(x) \, dx \, d\nu(y)$$

converges. It remains to apply Fubini's theorem and note that the two repeated integrals derived from (4_2) occur in (3).

1.6. Some questions. Let $G \subset R^m$ be an open set with a compact boundary B. For every $\nu \in \mathcal{E}'(B)$ we have then the generalized normal derivative $N^G \mathcal{U} \nu$ of the corresponding potential defined as a functional over $\mathcal{D}$. If there is a signed measure μ such that

$$\langle \varphi, N^G \mathcal{U} \nu \rangle = \int_{R^m} \varphi \, d\mu , \qquad \varphi \in \mathcal{D} ,$$

then we shall say, as usual, that $N^G \mathcal{U} \nu$ is a measure and write $N^G \mathcal{U} \nu = \mu$; in this case necessarily $\mu \in \mathcal{E}'(B)$ by remark 1.2. In general, however, $N^G \mathcal{U} \nu$ need not be a measure. We thus arrive naturally at the following

Question 1. Under which conditions on G can we assert that $N^G \mathcal{U} \nu \in \mathcal{C}'(B)$ for every $\nu \in \mathcal{C}'(B)$?

Our main objective in this paragraph is to answer this question in geometric terms connected with G . Before doing so we shall investigate the following simplified problem.

Question 2. Let $y \in B$ be a fixed point. What geometric conditions on G guarantee that $N^G \mathcal{U} \delta_y \in \mathcal{C}'(B)$?

In order to be able to answer this question we first introduce suitable terminology and establish several auxiliary results.

1.7. Definition. Let $S, M \subset R^m$. A point $y \in S$ will be termed a hit of S on M if for every $r > 0$ both

$$\mathcal{H}_1(\Omega_r(y) \cap S \cap M) > 0 \quad \text{and} \quad \mathcal{H}_1(\Omega_r(y) \cap (S \setminus M)) > 0 .$$

(In our applications S usually will be a straight line segment or a half-line.)

1.8. Lemma. Let $M \subset R^1$ be a Borel set and denote by χ_M its characteristic function on R^1 . If $a < b$, then

$$(5) \quad \sup\left\{ \int_a^b \chi_M(t)\psi'(t)dt \; ; \quad \psi \in \mathcal{D} , \quad |\psi| \leq 1, \quad \operatorname{spt} \psi \subset]a,b[\right\}$$

equals the total number of hits of $]a,b[$ on M (which is $+\infty$ if the set of all hits of $]a,b[$ on M is infinite).

Proof. Let q be the number of all hits of $]a,b[$ on M. Suppose first that $q < +\infty$ and let $a_1 < \dots < a_q$ be all the hits of $]a,b[$ on M. Then no $]a_j, a_{j+1}[$ can meet both M and $R^1 \setminus M$ in a set of positive linear measure. It follows that either M or $]a,b[\setminus M$ is $\mathcal{H}_1$-equivalent with

$$\bigcup_k \,]a_{2k-1}, a_{2k}[\,, \quad \text{where} \quad 1 \leqq k \,, \quad 2k \leqq q \,.$$

If $\psi \in \mathcal{D}$ and $\operatorname{spt} \psi \subset \,]a,b[$, then

$$\int_a^b \chi_M(t)\, \psi'(t)\,dt = \pm \sum_{j=1}^{q} (-1)^j \psi(a_j)$$

and the supremum (5) equals q.

Next suppose that the supremum (5) is finite. This means that the functional

$$L : \psi \longmapsto \int_a^b \chi_M(t)\, \psi'(t)\,dt$$

is bounded on the space $\mathcal{D}(]a,b[)$ of all infinitely differentiable functions ψ with $\operatorname{spt} \psi \subset \,]a,b[$ with respect to the norm $\|\psi\| = \sup_t |\psi(t)|$. Referring to the Hahn-Banach extension theorem and Riesz representation theorem we conclude that there is a function g of bounded variation on $]a,b[$ such that

$$\langle \psi, L \rangle = \int_a^b \psi\, dg = - \int_a^b g(t)\, \psi'(t)\,dt \,, \quad \psi \in \mathcal{D}(]a,b[) \,.$$

We see that the function $h = \chi_M + g$ satisfies

$$(6) \qquad \int_a^b h(t)\psi'(t)dt = 0$$

for all $\psi \in \mathcal{D}(]a,b[)$. It is known from elementary distribution theory that such an h must equal a constant c a.e. in $]a,b[$; we include here the proof of this fact for the sake of completeness.

Fix a $\varphi_o \in \mathcal{D}(]a,b[)$ with $\int_a^b \varphi_o(t)dt = 1$. If $\varphi \in \mathcal{D}(]a,b[)$ is arbitrarily chosen, then $\varphi - \varphi_o \int_a^b \varphi(t)dt$ has vanishing integral over $]a,b[$ and, consequently, there is a $\psi \in \mathcal{D}(]a,b[)$ with

$$\psi' = \varphi - \varphi_o \int_a^b \varphi(t)dt .$$

Employing (6) and writing $c = \int_a^b h(t)\,\varphi_o(t)dt$ we get $\int_a^b [h(t) - c]\,\varphi(t)dt = 0$. This is possible for all $\varphi \in \mathcal{D}(]a,b[)$ only if $h = c$ a.e. in $]a,b[$.

We have thus proved that the function $g_o = c - g$ equals χ_M a.e. in $]a,b[$. Since g_o has bounded variation and assumes both values 0 and 1 in every neighborhood of any hit of $]a,b[$ on M, we see that necessarily $q < +\infty$ and by the first part of the proof we are done.

<u>1.9. Notation.</u> If g is a locally integrable function in $]a,b[\subset R^1$, we define

$$\operatorname{varess}(g;]a,b[) = \sup\left\{ \int_a^b g(t)\,\psi'(t)\,dt \; ; \; \psi \in \mathcal{D}(]a,b[),\ |\psi| \leqq 1 \right\};$$

this quantity will be called the essential variation of g on $]a,b[$.

(The preceding lemma says that, for any Borel set $M \subset R^1$, the total number of hits of $]a,b[$ on M coincides with $\operatorname{varess}(\chi_M;]a,b[)$.)

<u>1.10. Lemma.</u> Let f be a bounded Baire function in R^m, fix $y \in R^m$ and put

(7) $$f_\theta(t) = f(y+t\theta), \quad t \in R^1, \quad \theta \in \Gamma.$$

If $0 \leqq a < b \leqq +\infty$, then

(8) $$\theta \longmapsto \operatorname{varess}(f_\theta ;]a,b[)$$

is a Baire function of the variable $\theta \in \Gamma$ and

$$\frac{1}{A} \int_\Gamma \operatorname{varess}(f_\theta ;]a,b[)\,d\mathcal{H}_{m-1}(\theta) = v(a,b,f),$$

where

$$v(a,b,f) = \sup\left\{ \int_{R^m} f(x)\,\operatorname{grad}\psi(x)\cdot\operatorname{grad} h_y(x)\,dx \; ; \; \psi \in \mathcal{D},\ |\psi| \leqq 1,\ \operatorname{spt}\psi \subset \{x \in R^m;\ a < |y-x| < b\} \right\}.$$

<u>Proof.</u> It is sufficient to investigate the case $y = 0$, $b < +\infty$ only. Applying the transformation formula in remark 1.4 to

$$g(x) = f(x) \operatorname{grad} \psi(x) \cdot \operatorname{grad} h_0(x) =$$

$$= -\frac{1}{A} \frac{1}{|x|^{m-1}} \operatorname{grad} \psi(x) \cdot \frac{x}{|x|} ,$$

we get with the notation from (7)

$$\int_{R^m} f(x) \operatorname{grad} \psi(x) \cdot \operatorname{grad} h_0(x) dx =$$

$$= -\frac{1}{A} \int_{\Gamma} \left(\int_a^b f_\theta(t) \psi'_\theta(t) dt \right) d\mathcal{H}_{m-1}(\theta) .$$

If we knew already that (8) is a Baire function, we would obtain

$$(9) \quad v(a,b,f) \leqq \frac{1}{A} \int_{\Gamma} \operatorname{varess}(f_\theta ;]a,b[) d\mathcal{H}_{m-1}(\theta) .$$

We have thus to prove that (8) is a Baire function and

$$(10) \qquad \frac{1}{A} \int_{\Gamma} \operatorname{varess}(f_\theta ;]a,b[) d\mathcal{H}_{m-1}(\theta) \leqq v(a,b,f) .$$

In order to do this we first adopt the following additional assumption on f :

(A) For every $\theta \in \Gamma$, f_θ is continuously differentiable on $]a,b[$ and

$$\sup \{ |f'_\theta(t)| ; \ \theta \in \Gamma , \ c < t < d \} \equiv K(c,d) < +\infty$$

whenever $a < c < d < b$.

For any positive integer N subdivide $]a,b[$ by means of the points

$$a_k \equiv a_k^N = a + k2^{-N}(b-a) , \quad 1 \leqq k < 2^N .$$

Consider $k < 2^N - 2$. Since

$$\theta \mapsto \operatorname{sign} (f_\theta (a_{k+1}) - f_\theta (a_k))$$

is a Baire function, there exist functions $\varphi_{ks} \in \mathcal{D}$ such that $|\varphi_{ks}| \leqq 1$ and

$$\lim_{s \to \infty} \varphi_{ks}(\theta) = \operatorname{sign} (f_\theta (a_{k+1}) - f_\theta (a_k)) \quad \text{for } \mathcal{H}_{m-1}\text{-}$$

$$\text{-almost every } \theta \in \Gamma .$$

Next we express the characteristic function of $]a_k,a_{k+1}[$ as $\lim_{s \to \infty} \rho_{ks}$, where ρ_{ks} are infinitely differentiable functions in R^1 with

$$\operatorname{spt} \rho_{ks} \subset]a_k,a_{k+1}[, \quad |\rho_{ks}| \leqq 1$$

and define

$$\psi_s(t\theta) = \sum_{k=1}^{2^N-2} \varphi_{ks}(\theta) \rho_{ks}(t) , \quad t \geqq 0 , \quad \theta \in \Gamma .$$

Then

$$\psi_s \in \mathcal{D} , \quad |\psi_s| \leqq 1, \quad \operatorname{spt} \psi_s \subset \{x \in R^m;\ a < |x| < b\} .$$

Consequently,

$$v(a,b,f) \geqq - \frac{1}{A} \int_\Gamma \Big(\int_a^b f_\theta (t) \psi'_{s\theta}(t)dt \Big) d\mathcal{H}_{m-1}(\theta) .$$

The sequence of functions

$$\theta \mapsto -\int_a^b f_\theta(t)\,\psi'_{s\theta}(t)dt = \sum_{k=1}^{2^N-2} \varphi_{ks}(\theta)\int_{a_k}^{a_{k+1}} \rho_{ks}(t)f'_\theta(t)dt$$

is dominated by $(b-a)K(a_1, a_{2^N-1})$ and converges, as $s \to \infty$, to

$$\sigma_N(\theta) = \sum_{k=1}^{2^N-2} |f_\theta(a_{k+1}) - f_\theta(a_k)|$$

for $\mathcal{H}_{m-1}$-every $\theta \in \Gamma$. Since f is now continuous on $]a,b[$, varess$(f;]a,b[)$ coincides with the ordinary total variation of f on $]a,b[$ and, consequently, $\sigma_N(\theta) \nearrow$ varess$(f_\theta\,;\,]a,b[)$ as $N \to \infty$. We see that, under the assumption (A), (8) is a Baire function and (10) holds.

Now we shall drop the additional assumption (A) and consider general f. For every positive integer N we fix a symmetric infinitely differentiable function ω^N in R^1 with

$$\text{spt }\omega^N \subset \left]-\tfrac{1}{N}, \tfrac{1}{N}\right[, \quad \int_{R^1} \omega^N(t)dt = 1$$

and define f^N so that, at the origin, $f^N(0) = 0$, while on $R^m \setminus \{0\}$ the values of f^N are determined in such a way that on the positive real axis $f^N_\theta \equiv (f^N)_\theta$ is equal to $f_\theta * \omega^N$ (= the convolution of f_θ and ω^N) for every $\theta \in \Gamma$. Let $a^N = a + \frac{1}{N}$, $b^N = b - \frac{1}{N}$, $\frac{2}{N} < b - a$. It follows from the first part of the proof that

$$(11)\quad \frac{1}{A}\int_{\Gamma} \operatorname{varess}(f_\theta^N ;]a^N,b^N[)\,d\mathcal{H}_{m-1}(\theta) = v(a^N,b^N,f^N) .$$

If ψ^N is obtained from ψ in the same way as f^N from f , then

$$\psi \in \mathcal{D}, \quad |\psi| \leqq 1, \quad \operatorname{spt} \psi \subset \{x \in R^m;\ a^N < |x| < b^N\}$$

imply

$$\psi^N \in \mathcal{D}, \quad |\psi^N| \leqq 1, \quad \operatorname{spt} \psi^N \subset \{x \in R^m;\ a < |x| < b\}$$

and

$$\int_{a^N}^{b^N} \psi_\theta'(t) f_\theta^N(t)\,dt = \int_a^b (\psi_\theta^N)'(t) f_\theta(t)\,dt .$$

Consequently,

$$(12)\qquad v(a^N,b^N,f^N) \leqq v(a,b,f) .$$

The same argument yields

$$(13)\qquad \operatorname{varess}(f_\theta^N ;]a^N,b^N[) \leqq \operatorname{varess}(f_\theta ;]a,b[) .$$

Since $f_\theta^N \to f_\theta$ $(N \to \infty)$ almost everywhere on $]0,+\infty[$, it follows from the definition of varess... that

$$(14)\qquad \liminf_{N\to\infty} \operatorname{varess}(f_\theta^N ;]a^N,b^N[) \geqq \operatorname{varess}(f_\theta ;]a,b[) ,$$

which together with (13) gives

$$\lim_{N\to\infty} \operatorname{varess}(f_\theta^N ;]a^N,b^N[) = \operatorname{varess}(f_\theta ;]a,b[) .$$

In particular, (8) is a Baire function. Combining (14) and Fatou's lemma with (11), (12) we arrive at (10), which completes the proof.

<u>1.11. Corollary.</u> Let $Q \subset R^m$ be a Borel set and fix $y \in R^m$ and $r > 0$. If $n_r^Q(\theta, y)$ denotes the number (possibly 0 or $+\infty$) of all hits of $\{y+t\theta ; 0 < t < r\}$ on Q, then

$$\theta \mapsto n_r^Q(\theta, y)$$

is a Baire function of the variable $\theta \in \Gamma$ and the quantity

$$v_r^Q(y) \equiv \frac{1}{A} \int_\Gamma n_r^Q(\theta, y) d\mathcal{H}_{m-1}(\theta)$$

equals

$$(15) \quad \sup\{\int_Q \text{grad}\ \psi(x) \cdot \text{grad}\ h_y(x); \quad \psi \in \mathcal{D}, \ |\psi| \leqq 1,$$
$$\text{spt}\ \psi \subset \Omega_r(y) \setminus \{y\}\}.$$

If φ is any continuously differentiable function with compact support such that

$$(16) \qquad \text{spt}\ \varphi \subset \Omega_r(y) \quad \text{and} \quad |\varphi| \leqq 1,$$

then

$$(17) \qquad |\int_Q \text{grad}\ \varphi(x) \cdot \text{grad}\ h_y(x) dx| \leqq 1 + v_r^Q(y).$$

<u>Proof.</u> If f denotes the characteristic function of Q on R^m, then (15) reduces to $v(0,r,f)$ from lemma 1.10 and it follows from 1.9 and 1.8 that $n_r^Q(\theta, y)$ equals $\text{varess}(f_\theta ;]0,r[)$. Lemma 1.10 gives the first part of the assertion. Suppose now that $v_r^Q(y) < +\infty$ and consider

a continuously differentiable function φ with (16). Using the notation from (7) and the transformation from remark 1.4 we get

$$(18) \quad \int_Q \operatorname{grad} \varphi(x) \cdot \operatorname{grad} h_y(x)dx = \int_{R^m} f(x) \operatorname{grad} \varphi(x) \cdot \operatorname{grad} h_y(x)dx =$$

$$= -\frac{1}{A} \int_\Gamma \Big(\int_0^r f_\theta(t)\, \varphi'_\theta(t)dt \Big) d\mathcal{H}_{m-1}(\theta) .$$

Fix $\theta \in \Gamma$ with $n_r^Q(\theta,y) = q < +\infty$ and let $y+t_j\theta$ be all the hits of $\{y+t\theta;\ 0 < t < r\}$ on Q, $0 < t_1 < \dots$ $\dots < t_q < r$. Then f_θ coincides a.e. in $[0,r]$ with the characteristic function of a finite union of disjoint closed intervals whose end-points are t_j and, possibly, 0 and r. Noting that φ_θ vanishes at r we get

$$\Big| \int_0^r f_\theta(t)\, \varphi'_\theta(t)dt \Big| \leqq 1 + n_r^Q(\theta,y) .$$

Since this is true for $\mathcal{H}_{m-1}$-almost every $\theta \in \Gamma$, (18) implies (17).

Now we are in position to answer the question 2 posed in 1.6. In what follows we write simply $v^Q(y)$, $n^Q(\theta,y)$ instead of $v_\infty^Q(y)$, $n_\infty^Q(\theta,y)$.

<u>1.12. Theorem.</u> Let $G \subset R^m$ be an open set with a compact boundary B, $y \in B$. Then $N^G \mathcal{U} \delta_y \in \mathcal{C}'(B)$ if and only if $v^G(y) < +\infty$. If $v^G(y) < +\infty$, then

$$v^G(y) \leqq \|N^G \mathcal{U} \delta_y\| \leqq 1 + v^G(y) .$$

Proof. It is sufficient to observe that $N^G \mathcal{U} \delta_y \in \mathcal{C}'(B)$ if and only if the supremum

$$(19) \qquad \sup\{< \varphi, N^G \mathcal{U} \delta_y >; \quad \varphi \in \mathcal{D}, \quad |\varphi| \leqq 1\}$$

is finite; if this is the case, then (19) equals the total variation of $N^G \mathcal{U} \delta_y$ on R^m and 1.11 gives

$$\|N^G \mathcal{U} \delta_y\| \geqq \sup\{\int_G \operatorname{grad} \psi(x) \cdot \operatorname{grad} h_y(x) dx ; \quad \psi \in \mathcal{D} ,$$

$$|\psi| \leqq 1, \quad \operatorname{spt} \psi \subset R^m \setminus \{y\}\} = v^G(y) .$$

Conversely, if $v^G(y) < +\infty$, then (17) in 1.11 yields

$$\sup\{< \varphi, N^G \mathcal{U} \delta_y >; \quad \varphi \in \mathcal{D}, \ |\varphi| \leqq 1\} \leqq 1 + v^G(y) .$$

Next we return to question 1 from 1.6. It is answered by the following result.

1.13. Theorem. Let $G \subset R^m$ be an open set with a compact boundary B . Then $N^G \mathcal{U} \nu \in \mathcal{C}'(B)$ for every $\nu \in \mathcal{C}'(B)$ if and only if the quantity

$$V^G \equiv \sup_{y \in B} v^G(y)$$

is finite. If $V^G < +\infty$, then the operator

$$N^G \mathcal{U} : \nu \longrightarrow N^G \mathcal{U} \nu$$

is bounded on $\mathcal{C}'(B)$,

$$(20) \qquad \|N^G \mathcal{U}\| \leqq 1 + V^G ,$$

and for every $\nu \in \mathcal{E}'(B)$ and any Borel set M

$$(21) \qquad N^G \mathcal{U} \nu(M) = \int_B N^G \mathcal{U} \delta_y(M) d\nu(y) .$$

Proof. With any $\varphi \in \mathcal{D}$ we associate the linear functional L_φ on $\mathcal{E}'(B)$ defined by

$$\langle \nu, L_\varphi \rangle = \langle \varphi, N^G \mathcal{U} \nu \rangle , \qquad \nu \in \mathcal{E}'(B) .$$

If $P_\varphi = G \cap \operatorname{spt} \varphi$ and $c_\varphi = \sup_x |\operatorname{grad} \varphi(x)|$, we have by (4_1) the estimate

$$|\langle \nu, L_\varphi \rangle| \leqq \int_G \int_B |\operatorname{grad} \varphi(x) \cdot \operatorname{grad} h_y(x)| \, dx \, d|\nu|(y) \leqq$$

$$\leqq c_\varphi \operatorname{diam}(P_\varphi \cup B) \|\nu\|$$

which shows that every functional L_φ is bounded on the Banach space $\mathcal{E}'(B)$ with the norm $\|\ldots\|$, $\|L_\varphi\| \leqq$ $\leqq c_\varphi \operatorname{diam}(P_\varphi \cup B)$. Let us denote $\mathcal{D}_1 = \{ \varphi \in \mathcal{D} ; |\varphi| \leqq 1 \}$. If $\nu \in \mathcal{E}'(B)$, then

$$(22) \quad N^G \mathcal{U} \nu \in \mathcal{E}'(B) \Longleftrightarrow \sup \{ \langle \varphi, N^G \mathcal{U} \nu \rangle ; \varphi \in \mathcal{D}_1 \} < +\infty .$$

Since $\langle \varphi, N^G \mathcal{U} \nu \rangle = \langle \nu, L_\varphi \rangle$, we observe that $N^G \mathcal{U} \nu \in$ $\in \mathcal{E}'(B)$ for every $\nu \in \mathcal{E}'(B)$ if and only if the class of functionals $\{L_\varphi\}_{\varphi \in \mathcal{D}_1}$ is pointwise bounded on $\mathcal{E}'(B)$. If this is so, then we conclude from the principle of uniform boundedness that

$$\sup_{\varphi \in \mathcal{D}_1} \|L_\varphi\| \equiv K < +\infty .$$

In particular, we get from theorem 1.12 for any $y \in B$

$$v^G(y) \leqq \sup_{\varphi \in \mathcal{D}_1} \langle \varphi, N^G \mathcal{U} \delta_y \rangle = \sup_{\varphi \in \mathcal{D}_1} \langle \delta_y, L\varphi \rangle \leqq K ,$$

so that $V^G \leqq K$.

Conversely, let $V^G < +\infty$. Using the observation 1.5 and the theorem 1.12 we get for any $\nu \in \mathcal{C}'(B)$

$$\sup_{\varphi \in \mathcal{D}_1} \langle \varphi, N^G \mathcal{U} \nu \rangle = \sup_{\varphi \in \mathcal{D}_1} \int_B \langle \varphi, N^G \mathcal{U} \delta_y \rangle d\nu(y) \leqq$$

$$\leqq (1 + V^G) \|\nu\| < +\infty .$$

In view of the equivalence (22) this means that $N^G \mathcal{U} \nu \in$ $\in \mathcal{C}'(B)$; besides that, $\|N^G \mathcal{U} \nu\| \leqq (1 + V^G) \|\nu\|$ and (20) is established. It is easy to see that the formula (3) now extends to any bounded Baire function φ ; applying this formula to the characteristic function of a Borel set M one gets (21).

<u>1.14. Remark.</u> If $C \subset R^m$, then the set of those $y \in R^m$ for which both

$$\mathcal{H}_m(\Omega_r(y) \cap C) > 0 \quad \text{and} \quad \mathcal{H}_m(\Omega_r(y) \setminus C) > 0$$

for every $r > 0$ is called the essential boundary of C and denoted by $\partial_e C$. Clearly, $\partial_e C$ is closed and contained in the boundary of C .

Let us now keep the notation from theorem 1.13. Assuming $V^G < +\infty$ we may consider the Neumann problem in the following formulation:

Given $\mu \in \mathcal{C}'(B)$, determine a $\nu \in \mathcal{C}'(B)$ with

(23) $$N^G \mathcal{U} \nu = \mu .$$

It turns out that it is sufficient to treat this problem only for those sets G whose boundary B coincides with $\partial_e G$. This follows from the following

<u>Observation.</u> Suppose that $G \subset R^m$ is an open set with a compact boundary B , $V^G < +\infty$, $\nu, \mu \in \mathcal{C}'(B)$ and

$$N^G \mathcal{U} \nu = \mu .$$

Let $B_e = \partial_e G$, put $G_e = G \cup (B \setminus B_e)$ and define for $\varkappa \in \mathcal{C}'(B)$

$$\varkappa_e(M) = \varkappa(M \cap B_e)$$

on Borel sets $M \subset R^m$, $\varkappa_o = \varkappa - \varkappa_e$.

Then G_e is open, its boundary coincides with $B_e = \partial_e G_e$, $V^{G_e} < +\infty$ and if $\mu_1 = N^G \mathcal{U} \nu_o$, then $\mu - \mu_1 \in \mathcal{C}'(B_e)$, $\nu_o = \mu_o$ and

$$N^{G_e} \mathcal{U} \nu_e = \mu - \mu_1 .$$

<u>Proof.</u> Clearly,

$$N^G \mathcal{U} \nu_e + N^G \mathcal{U} \nu_o = \mu_e + \mu_o .$$

Consider now $\varphi \in \mathcal{D}$ with $\operatorname{spt} \varphi \subset G_e$. Then

$$\langle \varphi, N^G \mathcal{U} \nu_e \rangle + \langle \varphi, N^G \mathcal{U} \nu_o \rangle = \langle \varphi, \mu_o \rangle .$$

Note that G_e is open, $G_e \supset G$ and $\mathcal{H}_m(G_e \setminus G) = 0$, so that $v^{G_e}(.) = v^G(.)$, $V^{G_e} \leqq V^G < +\infty$. Since $N^G \mathcal{U} \nu_e = N^{G_e} \mathcal{U} \nu_e$ and $\operatorname{spt} N^{G_e} \mathcal{U} \nu_e \subset B_e$, we have $< \varphi, N^G \mathcal{U} \nu_e > = 0$. Consequently, for our φ,

$$< \varphi, N^G \mathcal{U} \nu_o > = < \varphi, \mu_o > .$$

Now

$$< \varphi, N^G \mathcal{U} \nu_o > = < \varphi, N^{G_e} \mathcal{U} \nu_o > =$$

$$= \int_{G_e} \operatorname{grad} \varphi(x) \cdot \operatorname{grad} \mathcal{U}\nu_o(x) dx =$$

$$= \int_{R^m} \operatorname{grad} \varphi(x) \cdot \operatorname{grad} \mathcal{U}\nu_o(x) dx ,$$

because $\operatorname{spt} \varphi \subset G_e$. Since the integral

$$\int_{R^m \times B} \int \operatorname{grad} \varphi(x) \cdot \operatorname{grad} h_z(x) dx \, d\nu(z)$$

converges (compare the reasoning in 1.5) and, as we have seen in the example in 1.4,

$$\int_{R^m} \operatorname{grad} \varphi(x) \cdot \operatorname{grad} h_z(x) dx = \varphi(z) ,$$

we have

$$\int_{R^m} \operatorname{grad} \varphi(x) \cdot \operatorname{grad} \mathcal{U} \nu_o(x) =$$

$$= \int_B \left(\int_{R^m} \operatorname{grad} \varphi(x) \cdot \operatorname{grad} h_z(x) dx \right) d \nu_o(z) =$$

$$= \int \varphi \, d \nu_o ,$$

so that $\langle \varphi , \nu_o \rangle = \langle \varphi , \mu_o \rangle$. Noting that $|\varkappa_o|(R^m \setminus G_e) = 0$ for $\varkappa \in \mathcal{C}'(B)$ we see that

$$\nu_o = \mu_o$$

and, consequently,

$$N^{G_e} \mathcal{U} \nu_e = N^G \mathcal{U} \nu_e = \mu - N^G \mathcal{U} \mu_o .$$

It remains to note that, by theorem 1.13, $V^{G_e} < +\infty$ implies that $(\mu - \mu_1 =) \, N^{G_e} \mathcal{U} \nu_e \in \mathcal{C}'(B_e)$.

The above observation shows that the Neumann problem can always be reduced to another Neumann problem corresponding to a set whose boundary coincides with its essential boundary. Without loss of generality we may thus consider the equation (23) only for those sets G whose boundary is essential in the sense that it coincides with $\partial_e G$.

§ 2

Double layer potentials

If $G \subset R^m$ is an open set with a compact boundary B and $\nu \in \mathcal{C}'(B)$ then the generalized normal derivative $N^G \mathcal{U}\nu$ of the corresponding potential assumes at any test function $\varphi \in \mathcal{D}$ the value given by the integral

$$\int_{R^m} \left(\int_G \operatorname{grad} \varphi(x) \cdot \operatorname{grad} h_z(x) dx \right) d\nu(z) .$$

We see that the operator $\nu \mapsto N^G \mathcal{U}\nu$ is in a certain sense transposed to the operator associating with any $\varphi \in \mathcal{D}$ the integral

$$(1) \qquad \int_G \operatorname{grad} \varphi(x) \cdot \operatorname{grad} h_z(x) dx$$

considered as a function of the variable $z \in R^m$. The integral (1) has a good meaning for any Borel set $G \subset R^m$ and any continuously differentiable function φ with compact support and, as a function of z, has some useful properties which will be examined more closely in the present paragraph.

2.1. Lemma. Let $\mathcal{C}_0^{(1)} \equiv \mathcal{C}_0^{(1)}(R^m)$ denote the class of all continuously differentiable functions with compact support in R^m. Let $G \subset R^m$ be a Borel set, $z \in R^m$. If $\varphi_1, \varphi_2 \in \mathcal{C}_0^{(1)}$, $\varphi_1 = \varphi_2$ on $\partial_e G$ and $\varphi_1(z) = \varphi_2(z)$, then

$$\int_G \operatorname{grad} \varphi_1(x)\cdot \operatorname{grad} h_z(x)dx = \int_G \operatorname{grad} \varphi_2(x)\cdot \operatorname{grad} h_z(x)dx .$$

<u>Proof.</u> We shall first show that

$$(2) \qquad \int_G \operatorname{grad} \varphi(x)\cdot \operatorname{grad} h_z(x)dx = 0$$

for $\varphi \in \mathcal{E}_0^{(1)}$ vanishing in some neighborhood of the set $\{z\} \cup \partial_e G$. Let G_1 denote the set of those $x \in R^m$, for which there is an $r = r(x) > 0$ such that

$$(3) \qquad \mathcal{H}_m [\Omega_r(x) \setminus G] = 0 .$$

Clearly, G_1 is open, the boundary of G_1 is contained in $\partial_e G$ and, since $\operatorname{spt} \varphi \cap \partial_e G = \emptyset$,

$$\int_G \operatorname{grad} \varphi(x)\cdot \operatorname{grad} h_z(x)dx = \int_{G_1} \operatorname{grad} \varphi(x)\cdot \operatorname{grad} h_z(x)dx .$$

Fix $\psi \in \mathcal{D}$ such that $\psi = 1$ in some neighborhood of the compact set $\operatorname{spt} \varphi \cap \operatorname{cl} G_1 \subset G_1$ and $\psi = 0$ in some neighborhood of $\{z\} \cup (R^m \setminus G_1)$. If $\tilde{h} = \psi h_z$, then

$$\int_{G_1} \operatorname{grad} \varphi(x)\cdot \operatorname{grad} h_z(x)dx = \int_{G_1} \operatorname{grad} \varphi(x) \operatorname{grad} \tilde{h}(x)dx =$$

$$= \int_{R^m} \operatorname{grad} \varphi(x)\cdot \operatorname{grad} \tilde{h}(x)dx = - \int_{R^m} \varphi(x) \Delta \tilde{h}(x)dx = 0 ,$$

because $\varphi \Delta \tilde{h} = 0$ everywhere. Thus (2) is established in this case. Next we shall prove that (2) holds if $\varphi \in \mathcal{C}_0^{(1)}$ vanishes on $\{z\} \cup \partial_e G$. For every positive integer n we fix a continuously differentiable function ω_n on R^1 such that $|\omega_n'| \leqq 1$ on R^1 and

$$\omega_n(t) = \begin{cases} t - \frac{2}{n} & \text{for } t \geqq \frac{3}{n}, \\ 0 & \text{for } |t| \leqq \frac{1}{n}, \\ t + \frac{2}{n} & \text{for } t \leqq -\frac{3}{n}. \end{cases}$$

Then $\varphi_n(\cdot) = \omega_n(\varphi(\cdot))$ vanishes on $\{x \in R^m;\ |\varphi(x)| < \frac{1}{n}\}$ which is a neighborhood of $\{z\} \cup \partial_e G$, so that

$$\int_G \operatorname{grad} \varphi_n(x) \cdot \operatorname{grad} h_z(x)dx = 0 .$$

It follows from the definition of ω_n that, as $n \to \infty$, $\operatorname{grad} \varphi_n(x) \to \operatorname{grad} \varphi(x)$ for those x for which either $\varphi(x) \neq 0$ or else $\varphi(x) = 0$ and $|\operatorname{grad} \varphi(x)| = 0$ simultaneously. But the remaining set $\{x \in R^m;\ \varphi(x) = 0,\ |\operatorname{grad} \varphi(x)| \neq 0\}$ has zero Lebesgue measure (by the implicit function theorem, it can be covered by countably many smooth hypersurfaces). We see that $\operatorname{grad} \varphi_n \to \operatorname{grad} \varphi$ a.e. in R^m. Noting that all the functions $|\operatorname{grad} \varphi_n|$ are uniformly bounded and have supports in $\operatorname{spt} \varphi$ we get

$$\int_G \operatorname{grad} \varphi(x) \cdot \operatorname{grad} h_z(x)dx =$$

$$= \lim_{n \to \infty} \int_G \operatorname{grad} \varphi_n(x) \cdot \operatorname{grad} h_z(x)dx = 0 .$$

Thus (2) is proved for φ vanishing on $\{z\} \cup \partial_e G$. The rest is obvious.

The above lemma justifies the following

2.2. Definition. If $B \subset R^m$ is compact then $\mathcal{C}^{(1)}(B)$ denotes the class of all functions f on B for which there is a $\varphi \in \mathcal{C}_0^{(1)}(R^m)$ such that $f = \varphi$ on B . Suppose now that $G \subset R^m$ is a Borel set with a compact $\partial_e G$. For any $f \in \mathcal{C}^{(1)}(\partial_e G)$ and any $z \in R^m$ we choose a $\varphi^f \in \mathcal{C}_0^{(1)}$ in such a way that $\varphi^f = f$ on $\partial_e G$ and, in case $z \in R^m \setminus \partial_e G$, in addition $\varphi^f(z) = 0$, and define

$$W^G f(z) = \int_G \operatorname{grad} \varphi^f(x) \cdot \operatorname{grad} h_z(x) dx .$$

2.3. Remark. For physical reasons that will become clear later the function $W^G f : z \mapsto W^G f(z)$ is called the double layer potential with density f .

If $\tilde{G} \subset R^m$ is another Borel set such that the symmetric difference

$$\tilde{G} \dot{-} G = (\tilde{G} \setminus G) \cup (G \setminus \tilde{G})$$

has zero Lebesgue measure, then $\partial_e \tilde{G} = \partial_e G$ and $W^{\tilde{G}} f = W^G f$ on R^m for any $f \in \mathcal{C}^{(1)}(\partial_e G)$. It is easy to observe that, for any Borel set $G \subset R^m$, there is a Borel set $\tilde{G} \subset R^m$ with $\mathcal{H}_m(\tilde{G} \dot{-} G) = 0$ such that the boundary of $\tilde{G}$ coincides with $\partial_e G = \partial_e \tilde{G}$. For this purpose it is sufficient to put $\tilde{G} = G_1 \cup (G \setminus G_0)$, where G_0 is the set of those $x \in R^m$

for which there is an $r = r(x)$ such that

$$\mathcal{H}_m(\Omega_r(x) \cap G) = 0$$

and G_1 is the set of all $x \in R^m$ for which there is an $r = r(x)$ with (3). We may therefore without loss of generality restrict our investigation of double layer potentials $W^G f$ to those G for which $\partial_e G$ coincides with the whole boundary $\partial G = \mathrm{cl}\, G \cap \mathrm{cl}(R^m \setminus G)$.

Unless anything else is explicitly stated we always assume in the rest of this paragraph that $G \subset R^m$ is a Borel set with a compact boundary $B = \partial_e G$.

<u>2.4. Lemma.</u> If $f \in \mathcal{C}^{(1)}(B)$, then $W^G f$ is harmonic on $R^m \setminus B$ and $\lim_{|z| \to \infty} W^G f(z) = 0$.

<u>Proof.</u> If $y \in R^m \setminus B$ and $f \in \mathcal{C}^{(1)}(B)$, then there is a $\varphi^f \in \mathcal{E}_0^{(1)}$ such that $\varphi^f = f$ on B and $\varphi^f = 0$ on some open neighborhood $\mathcal{U}$ of y . Differentiating with respect to $z \in \mathcal{U}$ under the integral sign in

$$\int_{G \setminus \mathcal{U}} \mathrm{grad}\, \varphi^f(x) \cdot \mathrm{grad}\, h_z(x)dx = W^G f(z)$$

one gets immediately that $W^G f$ is harmonic in $\mathcal{U}$. It remains to note that, as $|z| \to \infty$, $|\mathrm{grad}\, h_z(x)| \to 0$ uniformly with respect to x in any bounded set.

<u>Remark.</u> For fixed $z \in R^m$,

$$f \longmapsto W^G f(z) \tag{4}$$

is a linear functional on $\mathcal{C}^{(1)}(B)$. We are now interested to know under which conditions (4) extends to a continuous linear functional on the Banach space $\mathcal{C}(B)$ of all continuous functions f on B equipped with the norm $\|f\| = \max |f|(B)$. The answer follows from 1.11 and we state it as a proposition.

2.5. Proposition. Fix $z \in R^m$. Then

(5) $$v^G(z) < +\infty$$

is a necessary and sufficient condition for extendability of (4) to a continuous linear functional on $\mathcal{C}(B)$. If (5) holds, then there is a uniquely determined signed measure $\lambda_z^G \in \mathcal{C}'(B)$ representing the functional (4) in the sense that

(6) $$W^G f(z) = \int_B f \, d\lambda_z^G , \quad f \in \mathcal{C}^{(1)}(B) ;$$

besides that,

(7) $$|W^G f(z)| \leq [1 + v^G(z)] \cdot \|f\| .$$

Proof. A necessary and sufficient condition for the functional (4) to be continuous with respect to the topology of uniform convergence consists clearly in the finiteness of

$$V(z) = \sup \{ W^G f(z); \; f \in \mathcal{C}^{(1)}(B), \; |f| \leq 1 \} .$$

If $V(z) < +\infty$ and $\psi \in \mathcal{D}$ is an arbitrary function with $|\psi| \leq 1$ vanishing in some neighborhood of z , then

$$\int_G \operatorname{grad} \psi(x) \cdot \operatorname{grad} h_z(x) dx \leqq V(z)$$

by the definition 2.2. Hence it follows by 1.11 that

$$v^G(z) \leqq V(z) .$$

Conversely, let $v^G(z) < +\infty$, consider an $f \in \mathcal{C}^{(1)}(B)$ and fix an arbitrary constant $k > \|f\|$. Then there is a $\varphi^f \in \mathcal{C}_0^{(1)}$ (vanishing in some neighborhood of z if $z \in R^m \setminus B$) such that $\varphi^f = f$ on B and $|\varphi^f| \leqq k$. According to 1.11 we have

$$|W^G f(z)| = \left| \int_G \operatorname{grad} \varphi^f(x) \cdot \operatorname{grad} h_z(x) dx \right| \leqq [1 + v^G(z)]\, k .$$

This proves the inequality (7) and finiteness of $V(z)$ which in turn implies the existence of a (uniquely determined) $\lambda_z^G \in \mathcal{C}'(B)$ with the property (6).

<u>Remark.</u> The above proposition shows that (5) is necessary for $W^G f(z)$ to be naturally defined for any $f \in \mathcal{C}(B)$. It permits us to extend the original definition of the double layer potential as follows.

<u>2.6. Definition.</u> If $v^G(z) < +\infty$, we define for any Baire function f on B

$$W^G f(z) = \int_B f \, d\lambda_z^G \tag{8}$$

provided the integral on the right-hand side converges.

(In particular, (8) is meaningful for each bounded Baire function f on B .)

In order to derive a geometric expression for $W^G f(z)$ it will be useful to adopt the following

2.7. Notation. We shall say that a vector $\theta \in \Gamma$ points into $M \subset R^m$ at $y \in R^m$ if there is a $\delta > 0$ such that

$$\mathcal{H}_1(\{y + t\theta;\ 0 < t < \delta\} \setminus M) = 0 ;$$

the set of all $\theta \in \Gamma$ that point into M at y will be denoted by M^y .

If $M \subset R^m$, $y \in R^m$ and $\theta \in \Gamma$, we put

$$s^M(\theta, y) = 1$$

provided θ points into M at y and, simultaneously, $-\theta$ points into $R^m \setminus M$ at y ; if $s^M(-\theta, y) = 1$ in the above described sense (i.e. $-\theta \in M^y$ and $\theta \in (R^m \setminus M)^y$), we let

$$s^M(\theta, y) = -1$$

and, finally, we define

$$s^M(\theta, y) = 0$$

in all remaining cases. Obviously, $s^M(\theta, y) \neq 0$ only if y is a hit of $\{y+t\theta;\ t \in R^1\}$ on M in the sense of 1.7.

With this notation we may formulate the following

2.8. Proposition. Assume (5). Then for any bounded Baire function f on B the sum

$$\sum\nolimits_f^G(\theta,z) \equiv \sum_{t>0} f(z+t\theta)s^G(\theta,z+t\theta)$$

is finite for $\mathcal{H}_{m-1}$-almost all $\theta \in \Gamma$, the function

$$\theta \longmapsto \sum\nolimits_f^G(\theta,z)$$

is $\mathcal{H}_{m-1}$-integrable over Γ and

$$(9) \qquad \int_{B\setminus\{z\}} f\,d\,\lambda_z^G = \frac{1}{A}\int_\Gamma \sum\nolimits_f^G(\theta,z)\,d\,\mathcal{H}_{m-1}(\theta)\ .$$

The set $G^z \subset \Gamma$ is $\mathcal{H}_{m-1}$-measurable and if $C = R^m \setminus G$, then

$$(10) \qquad \mathcal{H}_{m-1}(\Gamma \setminus G^z \setminus C^z) = 0\ ,$$

$$(11) \qquad \lambda_z^G(R^m \setminus \{z\}) = \begin{cases} -\frac{1}{A}\,\mathcal{H}_{m-1}(G^z) & \text{if } G \text{ is bounded,} \\ \frac{1}{A}\,\mathcal{H}_{m-1}(C^z) & \text{if } G \text{ is unbounded,} \end{cases}$$

$$(12) \qquad |\lambda_z^G|(R^m \setminus \{z\}) = v^G(z)\ .$$

If $z \in B$ then

$$(13) \qquad \lambda_z^G(\{z\}) = \frac{1}{A}\,\mathcal{H}_{m-1}(G^z)$$

and, for any bounded Baire function f on B,

$$(14) \qquad W^G f(z) + W^C f(z) = f(z)\ ,$$

while for $z \in R^m \setminus B$ we have for such an f

$$(15) \qquad W^G f(z) + W^C f(z) = 0$$

and for $\varphi \in \mathcal{E}_0^{(1)}$

$$(16) \quad \int_G \operatorname{grad} \varphi(x) \cdot \operatorname{grad} h_z(x) dx = W^G \varphi(z) + \frac{1}{A} \mathcal{H}_{m-1}(G^z) \varphi(z) .$$

Proof. Fix $\theta \in \Gamma$ with $n^G(\theta, z) < +\infty$ and consider the set

$$(17) \qquad S = \{ t > 0;\ s^G(\theta, z+t\theta) \neq 0 \} .$$

If

$$t_1 < \dots < t_p$$

are all its points, then

$$(18_1) \quad s^G(\theta, z+t_{j+1}\theta) = - s^G(\theta, z+t_j\theta), \quad 1 \leqq j < p ,$$

and

$$(18_2) \quad s^G(\theta, z+t_1\theta) = \begin{cases} -1 & \text{if } \theta \in G^z , \\ 1 & \text{if } \theta \notin G^z . \end{cases}$$

The set

$$G_\theta = \{ t > 0;\ z + t\theta \in G \}$$

is $\mathcal{H}_1$-equivalent with the union of finitely many disjoint closed intervals whose end-points are precisely the points of S to which one has to add 0 if $\theta \in G^z$ and $+\infty$ if G is unbounded. Let now $\varphi \in \mathcal{E}_0^{(1)}$ and put $\varphi_\theta(t) = \varphi(z+t\theta)$, $t > 0$. If χ stands for the characteristic function of G^z on Γ, we get

$$(19) \qquad - \int_G \varphi_\theta'(t) dt = \varphi(z) \chi(\theta) + \sum_\varphi^G (\theta, z) .$$

According to formula (18) established in §1 (where now $Q = G$, $y = z$, $r = +\infty$, f = characteristic function of G) this last integral, considered as a function of θ, is integrable $d\mathcal{H}_{m-1}(\theta)$ over Γ and

$$(20) \quad A \int_G \operatorname{grad} \varphi(x) \cdot \operatorname{grad} h_z(x)dx = -\int_\Gamma \Big(\int_{G_\theta} \varphi'_\theta(t)dt\Big) d\mathcal{H}_{m-1}(\theta) .$$

Let us denote by $\mathcal{B}$ the class of those bounded Baire functions φ on B, for which the corresponding function

$$\sigma_\varphi : \theta \longmapsto \varphi(z)\chi(\theta) + \sum_\varphi^G (\theta, z)$$

is integrable $d\mathcal{H}_{m-1}(\theta)$ over Γ. As we have just seen, $\mathcal{C}^{(1)}(B) \subset \mathcal{B}$. Since

$$\Big|\sum_\varphi^G (\theta, z)\Big| \leqq n^G(\theta, z) \cdot \sup |\varphi|(B)$$

and $\theta \longmapsto n^G(\theta, z)$ is integrable $d\mathcal{H}_{m-1}(\theta)$ over Γ, we infer from the Lebesgue dominated convergence theorem that $\mathcal{B}$ contains the limit of every uniformly bounded pointwise convergent sequence of its elements. Consequently, $\mathcal{B}$ contains all bounded Baire functions on B. If we let φ coincide with the characteristic function of $\{z\}$, we get $\sigma_\varphi = \chi$, so that G^z is $\mathcal{H}_{m-1}$-measurable. As we have seen above, for any $\varphi \in \mathcal{C}_0^{(1)}$ vanishing at z the integral (20) can be transformed into

$$A \int_{B \setminus \{z\}} \varphi \, d\lambda_z^G = \int_\Gamma \sum_\varphi^G (\theta, z) d\mathcal{H}_{m-1}(\theta) .$$

Repeating the previous argument one may extend the validity of this formula to all bounded Baire functions φ on B vanishing at z ; since the value of φ at z here is irrelevant, we see that (9) holds for any bounded Baire function f on B. Since

$$s^G(.,.) + s^C(.,.) = 0$$

by the definition described in 2.7, we get for such an f

$$(21) \qquad \int_{B\setminus\{z\}} f\,d\lambda_z^G + \int_{B\setminus\{z\}} f\,d\lambda_z^C = 0 .$$

Fix again an arbitrary $\theta \in \Gamma$ with $n^G(\theta,z) < +\infty$ and consider the corresponding set (17) with the points $t_1 < ..$ $.. < t_p$ and let $t_{p+1} = +\infty$ (so that $t_1 = +\infty$ if $S = \emptyset$). It is easily seen that $\mathcal{H}_1$-almost all points in $\{z+t\theta ;$ $0 < t < t_1\}$ are either contained in G (in which case $\theta \in$ $\in G^z$) or else contained in C (in which case $\theta \in C^z$); this proves (10), because $n^G(.,z) < +\infty$ almost everywhere $(\mathcal{H}_{m-1})$ on Γ. Suppose now that G is bounded. If $\theta \in G^z$, then necessarily $S \neq \emptyset$, $s^G(\theta, z+t_p\theta) = -1$ and, in view of (18_1), (18_2), we get for the constant function $\underline{1}$ identically equal to 1 on B the equality

$$\sum\nolimits_{\underline{1}}^{G}(\theta,z) = -1 .$$

If $\theta \in C^z$, then

$$\sum\nolimits_{\underline{1}}^{G}(\theta,z) = 0 .$$

Using (9) we obtain

$$\int_{B \setminus \{z\}} 1 \, d\lambda_z^G = - \frac{1}{A} \mathcal{H}_{m-1}(G^z) .$$

If G is unbounded, then C is bounded and by (21)

$$\int_{B \setminus \{z\}} 1 \, d\lambda_z^G = - \int_{B \setminus \{z\}} 1 \, d\lambda_z^C = \frac{1}{A} \mathcal{H}_{m-1}(C^z) .$$

Thus (11) is established.

According to the definition of the measure λ_z^G and the evaluation of the supremum (15) in 1.11 (where now $Q = G$, $y = z$, $r = +\infty$) we have

$$v^G(z) = \sup \left\{ \int_{R^m} \psi \, d\lambda_z^G ; \quad \psi \in \mathcal{D} , \quad |\psi| \leqq 1, \text{ spt } \psi \subset R^m \setminus \{z\} \right\}$$

which proves (12).

If $z \in B$ then, as we have seen in the beginning of this proof,

$$A \int_B \varphi \, d\lambda_z^G = \int_\Gamma \sigma_\varphi(\theta) d\mathcal{H}_{m-1}(\theta) =$$

$$= \varphi(z) \mathcal{H}_{m-1}(G^z) + \int_\Gamma \sum\nolimits_\varphi^G (\theta, z) d\mathcal{H}_{m-1}(\theta)$$

for $\varphi \in \mathcal{E}^{(1)}(B)$ and this formula clearly extends to all bounded Baire functions φ on B. Letting φ = characteristic function of $\{z\}$ we get (13). Noting that, by (10),

$$\mathcal{H}_{m-1}(G^z) + \mathcal{H}_{m-1}(C^z) = A$$

and using (21) we obtain (14).

If $z \in R^m \setminus B$, then $\lambda_z^G(\{z\}) = 0 = \lambda_z^C(\{z\})$ and (15) reduces to (21) while (19), (20), (9) yield (16) for any $\varphi \in \mathcal{C}_o^{(1)}$.

In what follows we shall need the following

<u>2.9. Lemma.</u> Let $z \in R^m$, $v^G(z) < +\infty$ and suppose that $E \subset \Gamma$ is measurable $(\mathcal{H}_{m-1})$ with $\mathcal{H}_{m-1}(E) > 0$. Put

$$\Lambda_E = \{z + t\theta;\ t > 0,\ \theta \in E\}.$$

Then

$$\lim_{r \to 0+} \frac{\mathcal{H}_m(\Omega_r(z) \cap \Lambda_E \cap G)}{\mathcal{H}_m(\Omega_r(z) \cap \Lambda_E)} = \frac{\mathcal{H}_{m-1}(G^z \cap E)}{\mathcal{H}_{m-1}(E)} .$$

In particular, the density of G at z defined by

$$d_G(z) = \lim_{r \to 0+} \frac{\mathcal{H}_m(\Omega_r(z) \cap G)}{\mathcal{H}_m(\Omega_r(z))} \tag{22}$$

exists and

$$A d_G(z) = \mathcal{H}_{m-1}(G^z) . \tag{23}$$

<u>Proof.</u> Let $G_r = \Omega_r(z) \cap G$ and put for $\theta \in \Gamma$

$$G_{r\theta} = \{t > 0;\ z+t\theta \in G_r\} .$$

Then the function

$$\theta \mapsto \mathcal{H}_1(G_{r\theta})$$

is $\mathcal{H}_{m-1}$-measurable on Γ (as one can easily see from remark 1.4) and, consequently, the set

$$G_r(z) = \{ \theta \in \Gamma \; ; \; \mathcal{H}_1(G_{r\theta}) = r \}$$

is measurable $(\mathcal{H}_{m-1})$. Elementary calculation yields

$$\mathcal{H}_m(G_r \cap \Lambda_E) \geqq \mathcal{H}_{m-1}(G_r(z) \cap E) r^m \cdot m^{-1} ,$$

$$\mathcal{H}_m(\Omega_r(z) \cap \Lambda_E) = \mathcal{H}_{m-1}(E) r^m \cdot m^{-1} .$$

Writing

$$d_r = \frac{\mathcal{H}_m(G_r \cap \Lambda_E)}{\mathcal{H}_m(\Omega_r(z) \cap \Lambda_E)} \qquad \left(\geq \frac{\mathcal{H}_{m-1}(G_r(z) \cap E)}{\mathcal{H}_{m-1}(E)} \right)$$

and noting that $G_r(z) \nearrow G^z$ as $r \searrow 0$ we get

$$(24) \qquad \liminf_{r \to 0+} d_r \geqq \frac{\mathcal{H}_{m-1}(G^z \cap E)}{\mathcal{H}_{m-1}(E)} .$$

Replacing G by $C = R^m \setminus G$ and observing that for $C_r =$ $= \Omega_r(z) \cap C$

$$\frac{\mathcal{H}_m(C_r \cap \Lambda_E)}{\mathcal{H}_m(\Omega_r(z) \cap \Lambda_E)} = 1 - d_r$$

we get

$$(25) \qquad \liminf_{r \to 0+} (1-d_r) \geqq \frac{\mathcal{H}_{m-1}(C^z \cap E)}{\mathcal{H}_{m-1}(E)} = 1 - \frac{\mathcal{H}_{m-1}(G^z \cap E)}{\mathcal{H}_{m-1}(E)} .$$

Combining (24) and (25) one obtains

$$\lim_{r \to 0+} d_r = \frac{\mathcal{H}_{m-1}(G^z \cap E)}{\mathcal{H}_{m-1}(E)} .$$

The existence of the density (22) and the equality (23)

follow from the special choice $E = \Gamma$.

Let us now adopt the following

<u>2.10. Notation.</u> For any Borel set $M \subset R^m$ put

$$P(M) = \sup_w \int_M \operatorname{div} w(x)dx ,$$

where $w = (w_1,\dots,w_m)$ ranges over all vector-valued functions with m components $w_j \in \mathcal{D}$ such that

$$|w|^2 \equiv \sum_{j=1}^{m} w_j^2 \leqq 1 .$$

This quantity $P(M)$ is called the perimeter of M.

It will be useful for us to examine certain relations between $v^G(.)$ and the perimeter $P(G)$.

<u>2.11. Proposition.</u> The function

$$z \mapsto v^G(z)$$

is lower semicontinuous on R^m and if we denote by

$$\operatorname{dist}(z,B) = \inf \{ |z-y| ; y \in B \}$$

the distance of z from B then, for any $z \in R^m \setminus B$,

$$v^G(z) \leqq \frac{1}{A} P(G) [\operatorname{dist}(z,B)]^{1-m} .$$

<u>Proof.</u> It follows from 1.11 that, for arbitrary $c < < v^G(z)$, there exists a $\psi \in \mathcal{D}$ with

(26) $\qquad |\psi| \leqq 1$ and $\operatorname{spt} \psi \subset R^m \setminus \{z\}$

such that

$$\int_G \operatorname{grad} \psi(x)\cdot \operatorname{grad} h_z(x)dx > c \ .$$

We have then for any y sufficiently close to z

$$\operatorname{spt} \psi \subset R^m \setminus \{y\}, \quad v^G(y) \geqq \int_G \operatorname{grad} \psi(x)\cdot \operatorname{grad} h_y(x)dx > c \ .$$

This proves the lower semicontinuity of $v^G(.)$.

Fix now a $z \in R^m \setminus B$ and an arbitrary positive number $\rho < \operatorname{dist}(z,B)$. Consider a $\psi \in \mathfrak{D}$ with the properties (26). Then there is a $\tilde{\psi} \in \mathfrak{D}$ vanishing on $\Omega_\rho(z)$ such that $|\tilde{\psi}| \leqq 1$ and $\tilde{\psi}$ coincides with ψ in some neighborhood of B . Defining

$$w(x) = A\tilde{\psi}(x) \operatorname{grad} h_z(x) \quad \text{for} \quad x \neq z \ ,$$

$$w(z) = 0 \quad (= \text{the zero vector in } R^m \ ),$$

we get

$$\operatorname{div} w(x) = A \operatorname{grad} \tilde{\psi}(x)\cdot \operatorname{grad} h_z(x) \ , \quad |w| \leqq \rho^{1-m} \ ,$$

whence it follows by lemma 2.1 and the definition of perimeter

$$A \int_G \operatorname{grad} \psi(x)\cdot \operatorname{grad} h_z(x)dx = A \int_G \operatorname{grad} \tilde{\psi}(x)\cdot \operatorname{grad} h_z(x)dx =$$

$$= \int_G \operatorname{div} w(x)dx \leqq \rho^{1-m} P(G) \ .$$

Since $\psi \in \mathfrak{D}$ was an arbitrary function with the properties (26), we conclude from 1.11 that

$$Av^G(z) \leqq \rho^{1-m} P(G) \ .$$

The proposition which we have just proved shows that $v^G(.) < +\infty$ on $R^m \setminus B$ if G has finite perimeter. We are now going to prove that the converse of this assertion is also true.

<u>2.12. Theorem.</u> If $z^1, \dots, z^{m+1} \in R^m$ are in general position (i.e. not situated in a single hyperplane) and

$$\sum_{j=1}^{m+1} v^G(z^j) < +\infty ,$$

then $P(G) < +\infty$.

<u>Proof.</u> Since we assume that $B = \partial G$ is compact, one of the sets G, $C = R^m \setminus G$ is necessarily bounded. Noting that $P(G) = P(C)$ and $v^G(.) = v^C(.)$ we may assume that G is bounded. Writing for $\psi \in \mathcal{D}$ and $\theta \in \Gamma$

$$\partial_\theta \psi = \theta \cdot \operatorname{grad} \psi$$

we are going to prove that

$$\sup \left\{ \int_G \partial_\theta \psi \ ; \ \psi \in \mathcal{D} , \ |\psi| \leqq 1 \right\} < +\infty$$

for every $\theta \in \Gamma$; hence $P(G) < +\infty$ easily follows. Fix $\theta \in \Gamma$ and denote by Π_j the hyperplane containing $\{z^k ; k \neq j\}$. Then

$$\bigcup_{j=1}^{m+1} (R^m \setminus \Pi_j) = R^m$$

and there are $\alpha_j \in \mathcal{D}$ such that $\Pi_j \cap \operatorname{spt} \alpha_j = \emptyset$ and such that $\alpha = \sum_{j=1}^{m+1} \alpha_j$ equals 1 in some neighborhood of the compact set cl G . We have then

$$\int_G \partial_\theta \psi(x)dx = \int_G \alpha(x) \partial_\theta \psi(x)dx$$

so that it suffices to verify for $j = 1,\ldots,m+1$

$$\sup\left\{ \int_G \alpha_j(x) \partial_\theta \psi(x)dx;\ |\psi| \leq 1,\ \psi \in \mathcal{D}\right\} < +\infty .$$

Consider $j = 1$. For $x \in \operatorname{spt} \alpha_1$ the vectors $x-z^2,\ldots$ $\ldots,x-z^{m+1}$ are linearly independent. Consequently,

$$\theta = \sum_{k=2}^{m+1} a_k(x) \frac{z^k-x}{|x-z^k|^m}$$

with infinitely differentiable coefficients $a_k(.)$ in some neighborhood of spt α_1 . Extending $a_k(.)$ arbitrarily to R^m we get

$$\int_G \alpha_1(x) \partial_\theta \psi(x)dx =$$

$$= \sum_{k=2}^{m+1} \int_G \alpha_1(x)a_k(x) \operatorname{grad} \psi(x)\cdot \frac{z^k-x}{|z^k-x|^m} dx .$$

Fix k and put $F(x) = \alpha_1(x)a_k(x)$. Then $F \in \mathcal{D}$ vanishes in the vinicity of z^k and

$$\int_G F(x) \operatorname{grad} \psi(x) \cdot \frac{z^k - x}{|z^k - x|^m} dx =$$

$$= \int_G \operatorname{grad} (F(x)\psi(x)) \cdot \frac{z^k - x}{|z^k - x|^m} dx -$$

$$- \int_G \psi(x) \operatorname{grad} F(x) \cdot \frac{z^k - x}{|z^k - x|^m} dx .$$

Clearly,

$$\left| \int_G \psi(x) \operatorname{grad} F(x) \cdot \frac{z^k - x}{|z^k - x|^m} dx \right| \leqq$$

$$\leqq \int_G |\operatorname{grad} F(x)| \cdot |z^k - x|^{1-m} dx < +\infty$$

and if $K = \max |F|$, then

$$\int_G \operatorname{grad} (F(x)\psi(x)) \cdot \frac{z^k - x}{|z^k - x|^m} dx =$$

$$= A \int_G \operatorname{grad} (F(x)\psi(x)) \cdot \operatorname{grad} h_{z^k}(x) dx \leqq AKv^G(z^k)$$

by 1.11.

2.13. Corollary. If $\mathcal{U} \subset R^m$ is open, we denote by $\mathbb{H}(\mathcal{U})$ the linear space of all harmonic functions in $\mathcal{U}$ endowed with the topology of uniform convergence on compact subsets of $\mathcal{U}$.

Consider now the operator

$$W^G : f \longmapsto W^G f \tag{27}$$

from $\mathcal{C}^{(1)}(B)$ into $\mathbb{H}(R^m \setminus B)$. Then $P(G) < +\infty$ is a necessary and sufficient condition for the operator (27) to be continuous with respect to the topology of uniform convergence in $\mathcal{C}^{(1)}(B)$.

Proof. If $P(G) < +\infty$ and $Q \subset R^m \setminus B$ is compact, then we get by proposition 2.11

$$V(Q) = \sup\{v^G(z);\ z \in Q\} \leqq A^{-1}P(G)\left[\operatorname{dist}(Q,B)\right]^{1-m} ,$$

where

$$\operatorname{dist}(Q,B) = \inf\{|z-y|;\ z \in Q,\ \ y \in B\} .$$

Since, by proposition 2.5,

$$\sup\{|W^G f(z)|;\ z \in Q\} \leqq [1 + V(Q)]\cdot\|f\|$$

for any $f \in \mathcal{C}^{(1)}(B)$, we see that the operator (27) is continuous from the topology of uniform convergence in $\mathcal{C}^{(1)}(B)$ into the topology of locally uniform convergence in $\mathbb{H}(R^m \setminus B)$. Conversely, if the operator W^G is continuous with respect to these topologies, then for every fixed $z \in R^m \setminus B$ the functional $f \longmapsto W^G f(z)$ is continuous with respect to the topology of uniform convergence in $\mathcal{C}^{(1)}(B)$, which implies $v^G(z) < +\infty$ by proposition 2.5. Since $v^G(.) < +\infty$ on $R^m \setminus B$, we conclude from theorem 2.12 that $P(G) < +\infty$.

2.14. Remark. The above results show that

$$P(G) < +\infty$$

is a natural condition for $W^G f(z)$ to be reasonably defined for sufficiently many functions f and sufficiently many points z .

We shall now recall several useful properties of sets with finite perimeter.

A vector $\theta \in \Gamma$ is termed the interior normal of G at $y \in R^m$ in Federer's sense, if the symmetric difference of G and the half-space $H = \{x \in R^m;\ (x-y)\cdot\theta > 0\}$ has m-dimensional density zero at y , i.e. $d_M(y) = 0$ for $M = G \dot{-} H$. If there is such a vector θ , then it is unique and we shall denote it $n^G(y) \equiv n(y)$; if there is no interior normal of G at y in this sense, we denote by $n(y)$ the zero vector in R^m . The vector-valued function $y \longmapsto n(y)$ is Borel measurable on R^m (cf. [Fe 1]). The set

$$\{y \in R^m;\ |n^G(y)| > 0\}$$

is called the reduced boundary of G and will be denoted by $\hat{\partial} G = \hat{B}$. An important property of sets with finite perimeter consists in validity of the following version of the divergence theorem (cf. [DG 1] , [Fe 2]).

<u>Divergence theorem.</u> If $P(G) < +\infty$, then $\mathcal{H}_{m-1}(\hat{\partial} G) < +\infty$ and for every vector-valued function $w = (w_1,\ldots,w_m)$ with components $w_j \in \mathcal{E}_0^{(1)}(R^m)$ the following Gauss-Green formula

$$\int_B w(y)\cdot n^G(y)\,d\,\mathcal{H}_{m-1}(y) = -\int_G \operatorname{div} w(x)\,dx$$

holds.

We shall occasionally also need the following relative isoperimetric inequality (cf. [Mi]; compare [FF] for isoperimetric inequalities concerning currents).

Isoperimetric lemma. Let $P(G) < +\infty$ and suppose that, for some $\varepsilon > 0$, $r > 0$, $z \in R^m$ the inequalities

$$\mathcal{H}_m(\Omega_r(z) \cap G) \geqq \varepsilon \mathcal{H}_m(\Omega_r(z)) ,$$

$$\mathcal{H}_m(\Omega_r(z) \setminus G) \geqq \varepsilon \mathcal{H}_m(\Omega_r(z))$$

hold. Then there is an $\alpha = \alpha(\varepsilon) > 0$ depending on ε only such that

$$\mathcal{H}_{m-1}(\Omega_r(z) \cap \hat{\partial} G) \geqq \alpha r^{m-1} .$$

The above formulated divergence theorem permits to give a new geometric interpretation to $v^G(z)$ and to the measure λ_z^G representing the value of double layer potentials at z.

2.15. Lemma. Let $P(G) < +\infty$. Then for each $z \in R^m$

$$v^G(z) = \int_B |n^G(y) \cdot \operatorname{grad} h_z(y)| \, d\mathcal{H}_{m-1}(y) \tag{28}$$

and if $v^G(z) < +\infty$ then for any Borel set $M \subset R^m$

$$\lambda_z^G(M \setminus \{z\}) = - \int_M n^G(y) \cdot \operatorname{grad} h_z(y) \, d\mathcal{H}_{m-1}(y) . \tag{29}$$

Proof. Fix $z \in R^m$ and consider an arbitrary $\psi \in \mathcal{E}_0^{(1)}$ vanishing in the vicinity of z.

Define

$$w(x) = \begin{cases} \psi(x) \text{ grad } h_z(x) & \text{if } x \neq z , \\ 0 \ (= \text{zero vector in } R^m) & \text{if } x = z . \end{cases}$$

Then $\operatorname{div} w = \operatorname{grad} \psi \cdot \operatorname{grad} h_z$ and, by the divergence theorem,

$$W^G \psi(z) = - \int_B n^G(y) \cdot w(y) d \mathcal{H}_{m-1}(y) .$$

We see that

(30) $$\int_{B \setminus \{z\}} \psi d \lambda_z^G = - \int_B \psi(y) n^G(y) \cdot \operatorname{grad} h_z(y) d \mathcal{H}_{m-1}(y) .$$

If we take here the supremum over all $\psi \in \mathcal{E}_o^{(1)}$ with the properties (26) we get

$$|\lambda_z^G|(R^m \setminus \{z\}) = \int_B |n^G(y) \cdot \operatorname{grad} h_z(y)| d \mathcal{H}_{m-1}(y)$$

which, in view of (12), is just the formula (28). Under the assumption $v^G(z) < +\infty$ the formula (30) extends to all bounded Baire functions ψ on R^m and yields (29) when applied to the characteristic function of a Borel set $M \subset R^m$.

Remark. Let $z \in R^m \setminus B$, $P(G) < +\infty$. We are now able to understand the reason for the physical terminology calling (for $m = 3$)

(31) $$W^G f(z) = - \int_{\hat{B}} f(y) n^G(y) \cdot \operatorname{grad} h_z(y) d \mathcal{H}_{m-1}(y)$$

the potential at z of the double layer with density f on $\hat{B}$. Indeed, if $y \neq z$ then $h_z(y)$ may be interpreted as the value at z of the potential of a unit charge situated at y. Suppose now that $n(y) \in \Gamma$ and we have a positive charge of magnitude $\frac{1}{2\varepsilon}$ at the point $y - \varepsilon n(y)$ and another negative charge of magnitude $-\frac{1}{2\varepsilon}$ at the point $y + \varepsilon n(y)$. For small $\varepsilon > 0$ the value at z of the potential of this configuration equals

$$\frac{1}{2\varepsilon}\left[h_z(y - \varepsilon n(y)) - h_z(y + \varepsilon n(y))\right] =$$

$$= - n(y)\cdot \text{grad}\ h_z(y) + o(1) .$$

Consequently, $- n(y)\cdot \text{grad}\ h_z(y)$ may be interpreted as the value at z of the potential of a dipole (= configuration of two infinite charges of opposite sign which are infinitely close to each other) at y with axis $n(y)$ and momentum (= product of magnitude and mutual distance of the charges) equal to 1. The integral (31) may then be given the meaning of the potential at z of a layer of continuously distributed dipoles in $\hat{B}$ whose axis at y is $n^G(y)$ and whose momentum has density f with respect to $\mathcal{H}_{m-1}$.

In what follows we shall need the following property of the function $v^G(.)$.

<u>2.16. Theorem.</u> Let

(32) $$V^G = \sup\{v^G(y);\ y \in B\}.$$

Then for any $z \in R^m$

(33) $$v^G(z) \leqq \frac{1}{2} + V^G .$$

<u>Proof.</u> The inequality (33) is obvious if $z \in B$ or $V^G = +\infty$. Let $V^G < +\infty$; then it follows easily from theorem 2.12 that $P(G) < +\infty$. Fix $z \in R^m \setminus B$ and $c < v^G(z)$. According to 1.11 there is a $\psi \in \mathcal{D}$ with the properties (26) such that

$$\int_G \operatorname{grad} \psi(x) \cdot \operatorname{grad} h_z(x)dx > c .$$

Consider first the case when $z \in R^m \setminus \operatorname{cl} G$ and denote by C_o the set of those $x \in R^m$ for which there is an $r = r(x) > 0$ with $\mathcal{H}_m(\Omega_r(x) \cap G) = 0$. Clearly, C_o is open and $\partial C_o \subset \partial_e G = B$, $\mathcal{H}_m(G \cap C_o) = 0$. The function

$$(34) \quad y \longmapsto \int_G \operatorname{grad} \psi(x) \cdot \operatorname{grad} h_y(x)dx =$$

$$= \int_{G \setminus C_o} \operatorname{grad} \psi(x) \cdot \operatorname{grad} h_y(x)dx$$

is harmonic in C_o, continuous on R^m and vanishes at infinity. Consider now an arbitrary $y \in B$ with $d_G(y) = \frac{1}{2}$. Then

$$\int_G \operatorname{grad} \psi(x) . \operatorname{grad} h_y(x)dx = W^G \psi(y) =$$

$$= \psi(y) d_G(y) + \int_{B \setminus \{y\}} \psi \, d \lambda_y^G \leqq d_G(y) + v^G(y) \leqq \frac{1}{2} + V^G$$

by proposition 2.8 and lemma 2.9. Since $P(G) < +\infty$ and $B = \partial_e G$, it follows from the isoperimetric lemma (cf. 2.14) that $\hat{\partial} G \subset \{y \in B;\ d_G(y) = \frac{1}{2}\}$ is dense in B. The continuous function (34) is therefore dominated by the same constant $\frac{1}{2} + V^G$ on the whole $B \supset \partial C_o$. Referring to the maximum principle for harmonic functions we conclude that $\frac{1}{2} + V^G$ is an upper bound for (34) on C_o and, in particular, at $z \in C_o$:

$$c < \int_G \operatorname{grad} \psi(x) \cdot \operatorname{grad} h_z(x)dx \leqq \frac{1}{2} + V^G .$$

Since $c < v^G(z)$ was arbitrarily chosen we get (33) in this case. In case $z \in \operatorname{cl} G \setminus B$ is the interior point of G the equality

$$\int_G \operatorname{grad} \psi(x) \cdot \operatorname{grad} h_z(x)dx = - \int_{R^m \setminus G} \operatorname{grad} \psi(x) \cdot \operatorname{grad} h_z(x)dx$$

(compare 2.8) permits to replace G by $C = R^m \setminus G$ and ψ by $-\psi$ in the above argument.

<u>2.17. Corollary.</u> For any $z \in R^m$ and $r > 0$

(35) $\mathcal{H}_{m-1}(\Omega_r(z) \cap \hat{\partial} G) \leqq Am(m+2)^m(\frac{1}{2} + V^G)r^{m-1}$.

<u>Proof.</u> Let $V^G < +\infty$ (so that $P(G) < +\infty$) and denote by $e^i \in R^m$ the point whose i-th coordinate equals $m+1$ and all remaining coordinates vanish. We have then for $\theta = (\theta_1, \ldots, \theta_m) \in \Gamma$

$$\sum_{i=1}^{m} |\theta \cdot e^i| \geqq (m+1) \sum_{i=1}^{m} |\theta_i| \geqq m + 1 ,$$

so that for $y \in \Omega_1(0)$

$$\left|\sum_{i=1}^{m} \theta(y-e^i)\right| \geqq \sum_{i=1}^{m} |\theta \cdot e^i| - \sum_{i=1}^{m} |\theta \cdot y| \geqq m+1-m = 1 ,$$

$$|y - e^i| \leqq m + 2 .$$

It is sufficient to prove (35) for $r = 1$ and $z = 0$, because V^G is invariant with respect to translations and dilations of G. Using the above inequalities and lemma 2.15 we get

$$\mathcal{H}_{m-1}(\Omega_1(0) \cap \hat{\partial} G) \leqq \sum_{i=1}^{m} \int_B |n^G(y) \cdot (y-e^i)| \, d\,\mathcal{H}_{m-1}(y) \leqq$$

$$\leqq (m+2)^m \sum_{i=1}^{m} \int_B \frac{|n^G(y) \cdot (y-e^i)|}{|y-e^i|^m} \, d\,\mathcal{H}_{m-1}(y) =$$

$$= (m+2)^m \, A \sum_{i=1}^{m} v^G(e^i) .$$

Employing (33) for $z = e^i$ $(i = 1,\ldots,m)$ we arrive at

$$\mathcal{H}_{m-1}(\Omega_1(0) \cap \hat{\partial} G) \leqq (m+2)^m A m(\tfrac{1}{2} + V^G) .$$

This corollary combined with the following lemma will permit us later to draw useful conclusions concerning potentials $\mathcal{U}_\nu$ of signed measures ν for which $d\nu = g|n^G| \, d\,\mathcal{H}_{m-1}$ with a bounded $\mathcal{H}_{m-1}$-measurable function g.

<u>2.18. Lemma.</u> Let ν be a signed measure with a compact support in R^m and suppose that for suitable constants $\lambda > m-2$ and $k > 0$

$$|\nu|(\Omega_r(z)) \leqq k\, r^{\lambda}$$

for all $z \in R^m$ and all $r > 0$. Then the function

$$(36) \qquad z \longrightarrow \int_{R^m} |h_z(x)|\, d\,|\nu|\,(x)$$

is locally bounded in R^m (and even bounded on the whole R^m if $m > 2$) and the potential $u = \mathcal{U}\nu$ satisfies the Hölder condition

$$|u(z^1) - u(z^2)| = o(|z^1-z^2|^{\varkappa}) \quad \text{as} \quad |z^1-z^2| \to 0+$$

for any $\varkappa$ such that $0 < \varkappa < \min(1, \lambda-m+2)$.

<u>Proof.</u> Let $\omega_z(\rho) := |\nu|(\Omega_\rho(z))$ and denote by $h(\rho)$ the value of h_o at the point $(\rho, 0, \dots, 0) \in R^m$. Then for $0 < R \leqq 1$

$$\int_{\Omega_R(z)} |h_z(x)|\, d\,|\nu|\,(x) = \int_0^R h(\rho)\, d\,\omega_z(\rho) \leqq$$

$$\leqq h(R)\,\omega_z(R) - \int_0^R \omega_z(\rho)\, dh(\rho) \leqq$$

$$\leqq kh(R)R^{\lambda} + kA^{-1} \int_0^R \rho^{1-m+\lambda}\, d\rho =$$

$$= kh(R)R^{\lambda} + kA^{-1}R^{2-m+\lambda} \cdot \frac{1}{2-m+\lambda} .$$

Since for any $\varepsilon \geqq 1$ with $\Omega_\varepsilon(z) \supset \operatorname{spt} \nu$

$$\int_{R^m \setminus \Omega_1(z)} |h_z(x)| \, d|\nu|(x) \leqq \|\nu\| \cdot \sup |h|([1, \varepsilon]) ,$$

we see that the function (36) is locally bounded (and even bounded if $m > 2$). Consider now arbitrary points $z^1, z^2 \in R^m$ with $0 < |z^1 - z^2| \leqq \frac{1}{2}$ and put $r = 2|z^1-z^2|$.

Then

$$|u(z^1) - u(z^2)| \leqq \int_{\Omega_r(z^1)} |h_{z^1}| \, d|\nu| + \int_{\Omega_r(z^1)} |h_{z^2}| \, d|\nu| +$$

$$+ \int_{R^m \setminus \Omega_r(z^1)} |h_{z^1} - h_{z^2}| \, d|\nu| \leqq \int_{\Omega_r(z^1)} |h_{z^1}| \, d|\nu| +$$

$$+ \int_{\Omega_{2r}(z^2)} |h_{z^2}| \, d|\nu| + \int_{R^m \setminus \Omega_r(z^1)} |h_{z^1} - h_{z^2}| \, d|\nu| .$$

If $x \in R^m \setminus \Omega_r(z^1)$, then

$$h_{z^1}(x) - h_{z^2}(x) = h_o(z^1-x) - h_o(z^2-x) =$$

$$= (z^1-z^2) \cdot \operatorname{grad} h_o(\xi - x) ,$$

where ξ is situated on the segment with end-points z^1, z^2, so that $|\xi - x| \geqq |z^1-x| - |z^1-z^2| \geqq \frac{1}{2} |z^1-x|$ and

$$|h_{z^1}(x) - h_{z^2}(x)| \leqq \frac{2^{m-1}}{A} |z^1-x|^{1-m} \cdot |z^1-z^2| .$$

Consequently,

$$\int_{R^m \setminus \Omega_r(z^1)} |h_{z^1} - h_{z^2}| \, d|\nu| \leqq \frac{2^{m-1}}{A} |z^1 - z^2| \int_r^\infty \rho^{1-m} \, d\omega_{z^1}(\rho) .$$

Noting that $\omega_{z^1}(\rho)$ remains constant for large ρ we get

$$\int_r^\infty \rho^{1-m} \, d\omega_{z^1}(\rho) \leqq (m-1) \int_r^\infty \omega_{z^1}(\rho) \, \rho^{-m} \, d\rho \leqq$$

$$\leqq (m-1) \Big[\int_r^1 \omega_{z^1}(\rho) \, \rho^{-m} \, d\rho + \|\nu\| \int_1^\infty \rho^{-m} \, d\rho \Big] \leqq$$

$$\leqq \|\nu\| + (m-1) k \int_r^1 \rho^{\lambda - m} \, d\rho .$$

Simple calculation yields

$$r \int_r^1 \rho^{\lambda - m} \, d\rho = o(r^{\varkappa}) \quad \text{as} \quad r \to 0+$$

for any $\varkappa$ with $0 < \varkappa < \min(1, \lambda - m + 2)$. Summarizing we obtain

$$|u(z^1) - u(z^2)| \leqq \int_{\Omega_r(z^1)} |h_{z^1}| \, d|\nu| +$$

$$+ \int_{\Omega_{2r}(z^2)} |h_{z^2}| \, d|\nu| + \int_{R^m \setminus \Omega_r(z^1)} |h_{z^1} - h_{z^2}| \, d|\nu| =$$

$$= o(r^{\varkappa}) \quad \text{as} \quad r = \frac{|z^1 - z^2|}{2} \to 0+ .$$

<u>2.19. Theorem.</u> Assume $P(G) < +\infty$ and define V^G by (32). Then $W^G f$ is bounded on $R^m \setminus B$ for every $f \in \mathcal{C}(B)$ if and only if

(37) $$V^G < +\infty .$$

If (37) holds, then for **any** $f \in \mathcal{C}(B)$ the corresponding double layer potential is uniformly continuous on each of the sets

$$G_i = \{x \in R^m \setminus B;\ d_G(x) = i\} \quad (i = 0,1)$$

and for $y \in \partial G_i$ the limit relation

(38) $$\lim_{\substack{x \to y \\ x \in G_i}} W^G f(x) = W^G f(y) - i f(y)$$

holds.

<u>Proof.</u> If $W^G f = \int f \, d\lambda_x^G$ is bounded as a function of the variable $x \in Q \subset R^m \setminus B$ for every $f \in \mathcal{C}(B)$ then, by the principle of uniform boundedness,

$$\sup_{x \in Q} \|\lambda_x^G\| = \Lambda < +\infty .$$

By (12) we get $v^G(x) \leqq \Lambda$ for $x \in Q$ and, in view of the lower semicontinuity of $v^G(.)$ (cf. 2.11), $v^G(.) \leqq \Lambda$ on cl Q. Applying this result to $Q = R^m \setminus B$ we obtain the necessity of (37). Conversely, assume (37). Using (33), 2.8 and 2.9 we get for any $f \in \mathcal{C}(B)$ and any $z \in R^m$

$$|W^G f(z)| \leqq (\tfrac{3}{2} + V^G) \cdot \|f\| .$$

We are now going to prove the limit relation (38) for fixed

$y \in \partial G_i$. Consider first the function $\mathbb{1}$ identically equal to 1 on B . According as G is bounded or not we get by 2.8 for $x \in R^m \setminus B$

$$W^G \mathbb{1}(x) = \lambda_x^G(R^m \setminus \{x\}) = \begin{cases} -d_G(x), \\ 1-d_G(x), \end{cases}$$

$$W^G \mathbb{1}(y) = d_G(y) + \lambda_y^G(R^m \setminus \{y\}) = \begin{cases} 0, \\ 1, \end{cases}$$

whence

$$\lim_{\substack{x \to y \\ x \in G_i}} W^G \mathbb{1}(x) = \begin{cases} -i = W^G \mathbb{1}(y) - i, \\ 1-i = W^G \mathbb{1}(y) - i, \end{cases}$$

and (38) is verified for $f = \mathbb{1}$ and, consequently, for all constant functions on B . It is therefore sufficient to verify (38) for $f \in \mathcal{C}(B)$ vanishing at y . For any positive integer k we may decompose such an f into $f_k + g_k = f$, where $f_k \in \mathcal{C}(B)$ vanishes in some neighborhood $\mathcal{U}_k$ of y and $|g_k| \leq \frac{1}{k}$. The function

$$W^G f_k(x) = - \int_{B \setminus \mathcal{U}_k} f_k(z) n^G(z) \operatorname{grad} h_x(z) d\mathcal{H}_{m-1}(z)$$

is then continuous (and even harmonic) in the vicinity of y while

$$|W^G g_k(.)| \leqq \left(\tfrac{3}{2} + V^G\right) \frac{1}{k} ,$$

as we have seen above. Consequently $W^G f$, being a uniform limit of $W^G f_k$ as $k \longrightarrow \infty$, is continuous at y and the proof is complete.

Example. Let $G = \Omega_1(0)$. Lemma 2.15 gives for $f \in \mathcal{E}(\Gamma)$, $x \in G$

$$W^G f(x) = \frac{1}{A} \int_\Gamma f(y) \frac{y(x-y)}{|x-y|^m} \, d\mathcal{H}_{m-1}(y) .$$

Since $y(y-x) = \frac{1}{2}(|y|^2 - |x|^2 + |y-x|^2)$, we get $W^G f(x) = - \frac{1}{2}(I_f(x) + U(x))$, where

$$I_f(x) = \frac{1}{A} \int_\Gamma f(y) \frac{1-|x|^2}{|x-y|^m} \, d\mathcal{H}_{m-1}(y)$$

is the so-called Poisson integral and

$$U(x) = \frac{1}{A} \int_\Gamma f(y) |x-y|^{2-m} \, d\mathcal{H}_{m-1}(y)$$

is continuous on R^m (cf. 2.18) and harmonic on G. Consequently, $I_f = - 2W^G f - U$ is harmonic on G and 2.19 yields for $z \in \Gamma$

$$\lim_{\substack{x \to z \\ x \in G}} I_f(x) = 2f(z) - 2W^G f(z) - U(z) .$$

Combining (13), (29) we get

$$W^G f(z) = \frac{1}{2} f(z) + \frac{1}{A} \int_\Gamma f(y) \frac{y(z-y)}{|z-y|^m} d\mathcal{H}_{m-1}(y) = \frac{1}{2}(f(z) - U(z)) ,$$

because $y(y-z) = \frac{1}{2}|y-z|^2$ (now $y, z \in \Gamma$). Hence

$$\lim_{\substack{x \to z \\ x \in G}} I_f(x) = f(z) ,$$

so that I_f provides explicit solution of the Dirichlet problem for $\Omega_1(0)$ corresponding to the boundary condition $f \in \mathcal{C}(\Gamma)$.

Remark. Let us denote by int M the interior of $M \subset R^m$. Then, in theorem 2.19,

$$G_1 = \text{int } G \quad \text{and} \quad G_o = \text{int}(R^m \setminus G) .$$

2.20. Proposition. Let $G \subset R^m$ be a Borel set with a compact boundary $B = \partial_e G$ and let

$$V^G = \sup_{y \in B} v^G(y) < +\infty .$$

Then for any $f \in \mathcal{C}(B)$ the restriction to B of the corresponding double layer potential $W^G f|_B$ is continuous on B, the operator

$$W_B^G : f \mapsto W^G f|_B \tag{39}$$

is bounded on $\mathcal{C}(B)$ and if we identify $\mathcal{C}'(B)$ with the dual space to $\mathcal{C}(B)$ and G is open, then the operator $N^G \mathcal{U}$ on $\mathcal{C}'(B)$ is dual to (39):

$$N^G \mathcal{U} = (W_B^G)' . \tag{40}$$

Proof. If $\varphi \in \mathcal{C}_o^{(1)}$ then, by definition 2.2, for $y \in B$

$$W^G \varphi(y) = \int_G \text{grad } \varphi(x) \text{ grad } h_y(x)\, dx$$

and the integral on the right-hand side is a continuous function of the variable y. It follows from (12), (13) that

$$|W^G f(y)| \leqq (1 + V^G)\|f\|$$

for any $f \in \mathcal{C}(B)$. Noting that for each $f \in \mathcal{C}(B)$ there is a sequence $\varphi_n \in \mathcal{D}$ such that $\varphi_n \to f$ uniformly on B

as $n \to \infty$ we conclude that $W_B^G \varphi_n \to W_B^G f$ $(n \to \infty)$ uniformly on B and, consequently, $W_B^G f \in \mathcal{C}(B)$, the operator (39) is bounded on $\mathcal{C}(B)$ and

$$\|W_B^G\| \leqq 1 + V^G .$$

Suppose now that G is open and $\nu \in \mathcal{C}'(B)$. We know from 1.5 and the definition 2.2 that for $\varphi \in \mathcal{D}$

$$\int_B W_B^G \varphi(y) d\nu(y) = \int_B \varphi \, dN^G \mathcal{U} \nu .$$

It follows from the above note that this equality extends to all $\varphi \in \mathcal{C}(B)$ and (40) is established.

2.21. Remark. If we are given a signed measure $\nu \in \mathcal{C}'(B)$, then the corresponding potential $\mathcal{U}\nu$ is defined and harmonic on $R^m \setminus B$, but it need not be defined at some points of B and it need not be bounded. We shall denote by $\mathcal{C}_c'(B)$ the subspace of those $\mu \in \mathcal{C}'(B)$ for which there exists a (finite) continuous function $\mathcal{U}_c \mu$ on R^m such that $\mathcal{U}_c \mu = \mathcal{U}\mu$ on $R^m \setminus B$. Assuming G open and $V^G < +\infty$ we are interested to know whether the operator $N^G \mathcal{U}$ preserves $\mathcal{C}_c'(B)$. We are going to prove that this is true if $\mathcal{H}_m(B) = 0$ and that, for $\mu \in \mathcal{C}_c'(B)$, the so-called Plemelj's exchange theorem ("Plemeljsche Umtauschsatz") holds which permits to calculate $\mathcal{U}_c N^G \mathcal{U}\mu$ from $\mathcal{U}_c \mu$ on B. For this purpose we shall need several auxiliary results.

In the rest of this paragraph we always suppose that G fulfils the assumptions of proposition 2.20.

2.22. Lemma. If $y \in B$ and $z \in R^m \setminus B$, then

$$\int_B h_z d\lambda_y^G = \int_B h_y d\lambda_z^G + d_G(z) h_y(z) .$$

Proof. We have

$$\lim_{|x| \to 0+} h_o(x) = \sup h_o(R^m) = +\infty ,$$

while

$$\lim_{|x| \to +\infty} h_o(x) = \inf h_o(R^m) = \begin{cases} 0 & \text{if } m > 2 , \\ -\infty & \text{if } m = 2 . \end{cases}$$

Fix a decreasing sequence $t_n < 1$ with

$$\lim_{n \to \infty} t_n = \inf h_o(R^m) \equiv I$$

and construct continuously differentiable functions ω_n on $]I,+\infty[$ in such a way that $\omega_n(t) = 0$ for $t < t_n$, ω_n remains constant on $]n+1,+\infty[$,

$$|\omega_n(t)| \leqq |t| , \quad |\omega_n'(t)| \leqq 1$$

and, besides that,

$$\lim_{n \to \infty} \omega_n(t) = t \quad \text{and} \quad \lim_{n \to \infty} \omega_n'(t) = 1 \quad \text{for} \quad t \in]I,+\infty[.$$

Next define

$$\varphi_n(x) = \begin{cases} \omega_n(h_z(x)) & \text{for } x \neq z , \\ \lim\limits_{t \to +\infty} \omega_n(t) & \text{for } x = z . \end{cases}$$

Then $\varphi_n \in \mathcal{C}_o^{(1)}$, $|\varphi_n| \leqq |h_z|$ and $\varphi_n \to h_z$ on $R^m \setminus \{z\}$ as $n \to \infty$. Noting that h_z is bounded on B, because $z \in R^m \setminus B$, we get

$$(42) \qquad \lim_{n\to\infty} \int_B \varphi_n d\lambda_y^G = \int_B h_z d\lambda_y^G .$$

By the definition of the measure λ_y^G we have

$$(43) \qquad \int_B \varphi_n d\lambda_y^G = \int_G \operatorname{grad} \varphi_n(x) \operatorname{grad} h_y(x) dx .$$

Further we have

$$|\operatorname{grad} \varphi_n| \leq |\operatorname{grad} h_z|, \quad \lim_{n\to\infty} \operatorname{grad} \varphi_n = \operatorname{grad} h_z \quad \text{on} \quad R^m \setminus \{z\}.$$

Since $|\operatorname{grad} h_z(x)|\cdot|\operatorname{grad} h_y(x)|$ is integrable and dominates $|\operatorname{grad} \varphi_n(x)|\cdot|\operatorname{grad} h_y(x)|$, we obtain

$$(44) \quad \lim_{n\to\infty} \int_G \operatorname{grad} \varphi_n(x) \operatorname{grad} h_y(x) dx = \int_G \operatorname{grad} h_z(x) \operatorname{grad} h_y(x) dx .$$

Using (42), (44) and making $n \to \infty$ in (43) we arrive at

$$(45) \quad \int_B h_z d\lambda_y^G = \int_G \operatorname{grad} h_z(x) \operatorname{grad} h_y(x) dx .$$

Further put

$$\psi_n(x) = \begin{cases} \omega_n(h_y(x)) & \text{for } x \neq y \\ \lim\limits_{t\to+\infty} \omega_n(t) & \text{for } x = y . \end{cases}$$

Again, $\psi_n \in \mathcal{E}_0^{(1)}$ and (for $z \in R^m \setminus B$) we get by (16), (22)

$$(46) \quad \int_B \psi_n d\lambda_z^G = \int_G \operatorname{grad} \psi_n(x) \operatorname{grad} h_z(x) - d_G(z)\, \psi_n(z) .$$

It follows from (29), 2.17 and 2.18 that

$$\int_B |h_y| d\, |\lambda_z^G| < +\infty .$$

Noting that $|\psi_n| \leq |h_y|$, $|\mathrm{grad}\ \psi_n| \leqq |\mathrm{grad}\ h_y|$ and making $n \to \infty$ in (46) we get now

$$\int_B h_y d\, \lambda_z^G = \int_G \mathrm{grad}\ h_y(x)\ \mathrm{grad}\ h_z(x) - d_G(z) h_y(z) ,$$

which together with (45) completes the proof.

Remarks. Let $\omega \geqq 0$ be a fixed symmetric continuous function with compact support in R^m . If u is locally integrable in R^m , then the convolution

$$u * \omega(x) = \int_{R^m} u(x-z)\omega(z)dz$$

is continuous; if u happens to have a compact support, then also $u * \omega$ has a compact support.

If now ν is a signed measure with compact support, then $\nu * \omega$ is defined as the functional on the space $\mathcal{C}_0(R^m)$ of all continuous functions with compact support by

$$\langle f, \nu * \omega \rangle = \langle f * \omega , \nu \rangle , \quad f \in \mathcal{C}_0(R^m) .$$

This functional is representable by a (uniquely determined) signed measure μ with compact support in the sense that

$$\langle f, \nu * \omega \rangle = \int f d\mu , \quad f \in \mathcal{C}_0(R^m) ;$$

μ is identified with $\nu * \omega$ and denoted by the same symbol: $\mu = \nu * \omega$.

Since the potential $\mathcal{U}\nu$ is locally integrable, one may form $(\mathcal{U}\nu) * \omega$ and compare it with $\mathcal{U}(\nu * \omega)$. It is easily seen that

(47) $$\mathcal{U}(\nu * \omega)(x) = (\mathcal{U}\nu) * \omega(x)$$

for all $x \in R^m$. Indeed, the integral

$$\iint |h_o(y-z-x)|\, \omega(z) dz d|\nu|(y)$$

is convergent, because the function

$$y \to \int |h_o(y-z-x)|\, \omega(z) dz$$

is bounded on the compact set spt ν. This justifies application of Fubini's theorem in the following calculations verifying (47):

$$\mathcal{U}(\nu * \omega)(x) = \int h_x * \omega \, d\nu = \iint h_x(y-z)\, \omega(z) dz d\nu(y) =$$

$$= \iint h_o(y-z-x)\, \omega(z) dz d\nu(y) \, ,$$

$$(\mathcal{U}\nu) * \omega(x) = \int \mathcal{U}\nu(x-z)\, \omega(z) dz =$$

$$= \iint h_{x-z}(y) d\nu(y)\, \omega(z) dz = \iint h_o(y-x+z) d\nu(y)\, \omega(-z) dz =$$

$$= \iint h_o(y-x-z)\, \omega(z) dz d\nu(y) \, .$$

Suppose now that for each n we have fixed a symmetric $\omega_n \geqq 0$ such that $\int_{R^m} \omega_n(x) dx = 1$ and

$$\lim_{n \to \infty} \operatorname{diam}(\{0\} \cup \operatorname{spt} \omega_n) = 0 \, .$$

Then for each continuous g we have $g * \omega_n \to g$ locally uniformly as $n \to \infty$ and hence it follows that, for each signed measure ν with compact support, $\nu * \omega_n \to \nu$ vaguely as $n \to \infty$ in the sense that, for any continuous g,

$$\lim_{n \to \infty} \int g\,d(\nu * \omega_n) = \int g\,d\nu .$$

These well-known elementary facts will be useful in the proof of the following result, which is sometimes called Plemelj's symmetry-rule ("Plemeljsche Symmetriegesetz").

2.23. Plemelj's exchange theorem. Suppose that G is open, fulfils the assumptions of proposition 2.20 and, in addition,

$$\mathcal{H}_m(B) = 0 .$$

Write simply $W_B = W_B^G$ (see (39)) and keep the notation introduced in 2.21; for $\mu \in \mathcal{C}_c'(B)$ let $\mathcal{U}_o\mu = \mathcal{U}_c\mu|_B$ denote the restriction of $\mathcal{U}_c\mu$ to B. If $\mu \in \mathcal{C}_c'(B)$, then $W_B'\mu$ $(= N^G\mathcal{U}\mu) \in \mathcal{C}_c'(B)$ and on B the equality

$$\mathcal{U}_o(W_B'\mu) = W_B(\mathcal{U}_o\mu)$$

holds.

Proof. For every n fix a continuous symmetric function $\omega_n \geqq 0$ with $\operatorname{spt} \omega_n \subset \Omega_{\frac{1}{n}}(0)$ such that

$$\int_{R^m} \omega_n(x)\,dx = 1 .$$

Let $\mu \in \mathcal{C}_c'(B)$, $\mu_n = \mu * \omega_n$, $\mathcal{U}_n\mu = (\mathcal{U}\mu) * \omega_n$.

Since $\mathcal{U}\mu = \mathcal{U}_c\mu$ almost everywhere $(\mathcal{H}_m)$ and $\mathcal{U}_c\mu$ is continuous, we conclude that

$$\lim_{n\to\infty} \mathcal{U}_n\mu = \mathcal{U}_o\mu$$

locally uniformly and, in particular, uniformly on B. Fix now $z \in R^m \setminus B$ and consider $\mathcal{U}_n\mu$ on B only. Then

$$\lim_{n\to\infty} W^G(\mathcal{U}_n\mu)(z) = W^G(\mathcal{U}_o\mu)(z) .$$

We have

$$(48)\quad W^G(\mathcal{U}_n\mu)(z) = \int_B \Big(\int_{R^m} h_x(y)\,d\mu_n(y)\Big) d\lambda_z^G(x) .$$

We conclude from 2.18, 2.17 and (29) that

$$y \mapsto \int_{R^m} |h_x(y)|\,d\,|\lambda_z^G|(x)$$

is bounded on spt μ_n and, consequently, the integral

$$\int_B \int_{R^m} |h_x(y)|\,d\,|\mu_n|(y)\,d\,|\lambda_z^G|(x)$$

converges. We are thus justified to apply Fubini's theorem and exchange the order of integrations in (48), so that

$$W^G(\mathcal{U}_n\mu)(z) = \int_{R^m} \Big(\int_B h_x(y)\,d\lambda_z^G(x)\Big) d\mu_n(y) .$$

Since

$$y \mapsto \int_B h_x(y)\,d\lambda_z^G(x) = \int_B h_y(x)\,d\lambda_z^G(x)$$

is a continuous function (cf. 2.18, 2.17 and (29)) and $\mu_n \to \mu$ vaguely as $n \to \infty$, we get

$$W^G(\mathcal{U}_o\mu)(z) = \int\limits_{R^m}\Big(\int\limits_B h_y d\,\lambda_z^G\Big)d\mu(y) .$$

On the other hand,

$$\mathcal{U}(W_B'\mu)(z) = \langle h_z, W_B'\mu\rangle = \int W_B h_z d\mu .$$

If $y \in B$, then lemma 2.22 yields

$$W_B^G h_z(y) = \int\limits_B h_z d\,\lambda_y^G = \int\limits_B h_y d\,\lambda_z^G + d_G(z)h_z(y) ,$$

whence

$$\mathcal{U}(W_B'\mu)(z) = \int\limits_{R^m}\Big(\int\limits_B h_y d\,\lambda_z^G\Big)d\mu(y) + d_G(z)\,\mathcal{U}\mu(z) =$$

$$= W^G(\mathcal{U}_o\mu)(z) + d_G(z)\,\mathcal{U}_c\mu(z) .$$

Employing theorem 2.19 we get

$$\lim_{\substack{z \to y\\ z \in R^m \setminus B}} \left[W^G(\mathcal{U}_o\mu)(z) + d_G(z)\,\mathcal{U}_c\mu(z)\right] = W_B(\mathcal{U}_o\mu)(y) .$$

We see that $\mathcal{U}(W_B'\mu)$ has a limit at any $y \in B$, so that $W_B'\mu \in \mathcal{C}_c'(B)$, and, besides that,

$$\mathcal{U}_o(W_B'\mu)(y) = W_B(\mathcal{U}_o\mu)(y) .$$

2.24. Remark. Let us keep the assumptions of proposition 2.20 and put $C = R^m \setminus G$. We know from theorem 2.19 that, for $y \in B \cap \mathrm{cl}(\mathrm{int}\ C)$ and $f \in \mathcal{C}(B)$,

$$\lim_{\substack{x \to y\\ x \in \mathrm{int}\ C}} W^G f(x) = W^G f(y) .$$

If $B \subset \mathrm{cl}(\mathrm{int}\ C)$, $g \in \mathcal{C}(B)$ is prescribed on B and we wish to determine the solution of the Dirichlet problem for int C and the boundary condition g in the form $W^G f$ with an unknown $f \in \mathcal{C}(B)$, we arrive at the equation

(49) $$W_B^G f = g .$$

If G is open, then the generalized Neumann problem for G and the prescribed boundary condition $\mu \in \mathcal{C}'(B)$ (cf. remark 1.14) consists in determining a $\nu \in \mathcal{C}'(B)$ satisfying

(50) $$(W_B^G)' \nu = \mu$$

(cf. proposition 2.20). The dual equations (49), (50) may conveniently be written in the form

$$[\alpha I + (W_B^G - \alpha I)] f = g ,$$

$$[\alpha I + (W_B^G - \alpha I)]' \nu = \mu ,$$

where I stands for the identity operator on $\mathcal{C}(B)$ and $\alpha \in R^1$ is a parameter. We shall see later that, for several reasons, the optimal choice of the parameter is $\alpha = \frac{1}{2}$.

2.25. Proposition. Under the assumptions of 2.20

$$\| W_B^G - \alpha I \| = |\alpha - \tfrac{1}{2}| + V^G ;$$

consequently,

$$\min_{\alpha \in R^1} \| W_B^G - \alpha I \| = \| W_B^G - \tfrac{1}{2} I \| = V^G .$$

Proof. We have

$$\| W_B^G - \alpha I \| = \sup_{\substack{f \in \mathcal{C}(B) \\ \|f\| \leq 1}} \ \sup_{y \in B} \ (W_B^G - \alpha I) f(y) .$$

Since $B = \partial_e G$, the reduced boundary $\hat{B}$ is dense in B by the isoperimetric lemma (cf. 2.14); noting that $d_G(y) = \frac{1}{2}$ whenever $y \in \hat{B}$, we get for $f \in \mathcal{C}(B)$ by 2.8 and (23)

$$\sup_{y \in B} (W_B^G - \alpha I)f(y) = \sup_{y \in \hat{B}} \left[(\tfrac{1}{2} - \alpha)f(y) + \int_{B \setminus \{y\}} f \, d\lambda_y^G\right],$$

whence we conclude by (12)

$$\|W_B^G - \alpha I\| = \sup_{y \in \hat{B}} \left[|\tfrac{1}{2} - \alpha| + |\lambda_y^G|(B \setminus \{y\})\right] =$$

$$= |\tfrac{1}{2} - \alpha| + \sup_{y \in \hat{B}} v^G(y) .$$

In view of the lower-semicontinuity of $v^G(\cdot)$ (cf. 2.11) we have

$$\sup_{y \in \hat{B}} v^G(y) = V^G$$

and the proof is complete.

Remark. In the next paragraphs we shall be concerned with more detailed investigation of the operator $W_B^G - \frac{1}{2} I = \frac{1}{2}(2W_B^G - I)$ on $\mathcal{C}(B)$; the operator

$$T^G = 2W_B^G - I \tag{51}$$

is called Neumann's operator of the arithmetical mean.

§ 3

Contractivity of Neumann's operator

Unless anything else is explicitly stated we assume in this paragraph that $G \subset R^m$ is a Borel set with a compact boundary $B = \partial_e G \neq \emptyset$ such that $V^G < +\infty$ (cf. (32) in §2 for notation). We put $C = R^m \setminus G$. In view of the compactness of B one of the sets G, C is bounded; for definiteness, we assume that C is bounded. We define Neumann's operator $T^G \equiv T$ by (51) in §2, so that $W_B^G = \frac{1}{2}(I+T)$ and, for $z \in B$ and $f \in \mathcal{C}(B)$,

(1) $$Tf(z) = \int_B f \, d\tau_z ,$$

where $\tau_z \equiv \tau_z^G \in \mathcal{C}'(B)$ is given by

(2) $$\tau_z^G = 2\lambda_z^G - \delta_z$$

(here δ_z is the Dirac measure at z and λ_z^G is defined by (6) in proposition 2.5); it follows from 2.8, 2.9 and 2.15 that

(3) $$d\tau_z^G(y) = [2d_G(z)-1]\, d\delta_z(y) - 2n^G(y)\cdot \text{grad } h_z(y)\, d\mathcal{H}_{m-1}(y) ,$$

where $n^G(\cdot)$ is the vector-valued function defined in 2.14. In terms of T the equations (49) and (50) in §2 (which, as we observed in remark 2.24, are connected with the Dirichlet and the Neumann problem, respectively) may be transformed into the form

(4) $$(I + T)f = 2g ,$$

(5) $(I + T)'\nu = 2\mu$.

3.1. Theorem. $\|T\| \leqq 1$ if and only if cl C is convex and $\mathcal{H}_m(\mathrm{cl}\, C \setminus C) = 0$.

Proof. Let $z \in \hat{B}$ and let $n(z) = n^G(z)$ denote the interior normal of G at z in Federer's sense (cf. 2.14). We shall first prove the following

Lemma. G^z is equivalent $(\mathcal{H}_{m-1})$ with $\{\theta \in \Gamma;\ \theta \cdot n(z) > 0\}$ and C^z is equivalent with $\{\theta \in \Gamma;\ \theta \cdot n(z) < 0\}$ (cf. 2.7 for notation).

Indeed, if we put $M = \{\theta \in \Gamma;\ \theta \cdot n(z) < 0\}$, $\Lambda_M = \{x \in R^m;\ (x-z)\cdot n(z) < 0\}$ then, by the definition of Federer's normal,

$$\lim_{r \to 0+} \frac{\mathcal{H}_m[\Omega_r(z) \cap \Lambda_M \cap G]}{\mathcal{H}_m[\Omega_r(z) \cap \Lambda_M]} = 0$$

and lemma 2.9 yields $\mathcal{H}_{m-1}(G^z \cap M) = 0$. Similar reasoning gives $\mathcal{H}_{m-1}(C^z \setminus M) = 0$ and, in view of (10) in §2, the lemma is established.

Suppose now that $\|T\| \leqq 1$ and denote $D(z) = \{x \in R^m;\ (x-z)\cdot n(z) \leqq 0\}$. We assert that $B \subset D(z)$. In the opposite case the open half-space $R^m \setminus D(z)$ meets C in a set of positive $\mathcal{H}_m$-measure. Since $\mathcal{H}_{m-1}$-almost all vectors in $\Gamma_+ = \{\theta \in \Gamma;\ \theta \cdot n(z) > 0\}$ point into G at z we conclude that the set of those $\theta \in \Gamma_+$, for which the half-line $\{z+t\theta;\ t > 0\}$ meets both G and C in a set of positive $\mathcal{H}_1$-measure, has positive $\mathcal{H}_{m-1}$-measure; for such θ necessarily $n_\infty^G(\theta, z) \geqq 1$. Since C is bounded and $\mathcal{H}_{m-1}$-almost all vectors in $\Gamma_- = \{\theta \in \Gamma\, ;\, \theta \cdot n(z) < 0\}$ belong to C^z, we conclude that $n_\infty^G(\theta, z) \geq 1$ for $\mathcal{H}_{m-1}$-almost all $\theta \in \Gamma_-$. Consequently,

$$v^G(z) = \frac{1}{A} \int_{\Gamma_+} n_\infty^G(\theta, z) d\mathcal{H}_{m-1}(\theta) + \frac{1}{A} \int_{\Gamma_-} n_\infty^G(\theta, z) d\mathcal{H}_{m-1}(\theta) >$$

$$> \frac{1}{A} \int_{\Gamma_-} n_\infty^G(\theta, z) d\mathcal{H}_{m-1}(\theta) = \frac{1}{2} .$$

According to (12) in §2 and (2), (1) this would mean that $\|\tau_z^G\| \geq |\tau_z^G|(R^m \setminus \{z\}) = 2v^G(z) > 1$, so that $\|T\| > 1$ - a contradiction.

Consequently, $\text{cl } C \subset D(z)$ for any $z \in \hat{B}$, so that

$$\text{cl } C \subset \bigcap_{z \in \hat{B}} D(z) = C_1 .$$

Since $B \subset \partial C_1$ and $\hat{B}$ is dense in $B = \partial C = \partial_e G$ by the isoperimetric lemma (cf. 2.14), we conclude that $B \subset \partial C_1$, so that $\text{cl } C = C_1$ is convex; clearly, $\text{cl } C \setminus C \subset B \subset \partial C_1$ has $\mathcal{H}_m$-measure zero.

Conversely, let $\text{cl } C$ be convex and $\mathcal{H}_m(\text{cl } C \setminus C) = 0$. Consider an arbitrary $z \in B$. Then $d_G(z) \geq \frac{1}{2}$ and for any y with $n^G(y) \neq 0$ the half-space $D(y) = \{x \in R^m;\ (x-y)\cdot n^G(y) \leq \leq 0\}$ contains z, so that

$$n^G(y) \cdot \text{grad } h_z(y) \leq 0 .$$

Consequently, $\tau_z^G \geq 0$ by (1), (3). Employing (1), (3), 2.15, 2.8, 2.9 we get for any $f \in \mathcal{C}(B)$ with $\|f\| \leq 1$

$$|Tf(z)| \leq [2d_G(z)-1]\, |f(z)| -$$

$$- \int_{B \setminus \{z\}} n^G(y) \cdot \operatorname{grad} h_z(y) \, |f(y)| \, d\mathcal{H}_{m-1}(y) \leqq$$

$$\leqq [2d_G(z)-1] + \lambda_z^G(R^m \setminus \{z\}) = 2d_G(z) - 1 + d_C(z) =$$

$$= d_G(z) \leqq 1 \text{ , so that } \|T\| \leqq 1 \text{ .}$$

3.2. Remark. We are now going to investigate more closely the Neumann operator T^G for the case that cl C is convex and $\mathcal{H}_m(\text{cl } C \setminus C) = 0$. Since we assume $B = \partial_e G \neq \emptyset$, T^G is not affected if we replace G by $R^m \setminus \text{cl } C$ and C by cl C . We shall thus assume in the rest of this paragraph without any loss of generality that $C = \text{cl } C$ is a convex body and $G = R^m \setminus \text{cl } C$ is open.

It follows from (2) and 2.8 that

(6) $$\tau_z(R^m) = 1 \quad (z \in B);$$

in other words,

(7) $$T \, \underline{1} = \underline{1}$$

for the function $\underline{1}$ identically equal to 1 on B . If $\mathcal{K} = \mathcal{K}(B)$ denotes the space of all constant functions on B we have thus

(8) $$T(\mathcal{K}) \subset \mathcal{K} \text{ .}$$

We have observed in the course of the proof of theorem 3.1 that

(9) $$z \in B \Longrightarrow \tau_z \geqq 0 \text{ ,}$$

so that the operator T is positive on $\mathcal{C}(B)$ (in the sense that $Tf \geqq 0$ whenever $f \in \mathcal{C}(B)$ is non-negative on B).
It follows from 3.1 and (7) that

$\|T\| = 1$.

In view of (8) we may consider T on $\mathcal{C}(B)_{/\mathcal{K}}$; adopting the usual definition of the norm in the factor-space we are interested to know under which conditions on C the resulting operator is contractive ($\Leftrightarrow$ has norm strictly less than 1), which is important in connection with the equations (4), (5).

If $Q \subset B$ is compact, we put for $f \in \mathcal{C}(B)$

$$\operatorname{osc} f(Q) = \max f(Q) - \min f(Q) ;$$

in this notation the norm of the class determined by f in $\mathcal{C}(B)_{/\mathcal{K}}$ equals $\frac{1}{2}\operatorname{osc} f(B)$. Accordingly, T is contractive on $\mathcal{C}(B)_{/\mathcal{K}}$ if and only if there is a $q \in [0,1[$ such that

$$(10) \qquad f \in \mathcal{C}(B) \Longrightarrow \operatorname{osc} Tf \leqq q \operatorname{osc} f ,$$

where we write simply $\operatorname{osc} f = \operatorname{osc} f(B)$; we are now going to characterize those convex bodies C for which this is true. For this purpose we shall need two auxiliary results.

<u>3.3.</u> <u>Lemma.</u> We shall denote by $Q_x(C)$ the smallest closed cone of vertex $x \in R^m$ containing C ; thus $Q_x(C) \equiv Q_x$ is the union of all the half-lines

$$H_x(\theta) = \{x + t\theta ;\ t \geqq 0\}$$

for which there are points $z^n \in C \setminus \{x\}$ with

$$\lim_{n\to\infty} \frac{z^n - x}{|z^n - x|} = \theta .$$

If $z^1, z^2 \in B = \partial C$ and $Q_{z^1} \cap Q_{z^2} \neq C$, then there exist constants $\rho > 0$ and $q \in]0,1[$ such that

$$\|\tau_{x^1} - \tau_{x^2}\| \leqq 2q$$

whenever $x^i \in \Omega_\rho(z^i) \cap B$, $i = 1,2$.

<u>Proof.</u> Fix $z \in B \cap \operatorname{int} Q_{z^1} \cap \operatorname{int} Q_{z^2}$ and choose $r > 0$ in such a way that

$$\operatorname{cl} \Omega_r(z) \subset \operatorname{int} Q_{z^1} \cap \operatorname{int} Q_{z^2} .$$

Since the function

$$n : y \longmapsto n^G(y)$$

is Borel measurable on the set $\hat{B} \cap \operatorname{cl} \Omega_r(z)$ which has positive $\mathcal{H}_{m-1}$-measure (cf. 2.14) we conclude from Luzin's theorem that there is a compact $Q \subset \hat{B} \cap \operatorname{cl} \Omega_r(z)$ such that the restriction of n to Q is continuous. Let us observe that, for each $y \in Q$, the whole segment $\{z^i + t(y-z^i);\ 0 < t < 1\}$ is contained in $\operatorname{int} C$, C has a unique supporting hyperplane at y and both the quantities

$$n(y)\cdot(y-z^1) , \quad n(y)\cdot(y-z^2)$$

are different from zero and of the same sign. Since the function

$$(y,x) \longmapsto n(y)\cdot(y-x)$$

is continuous on $Q \times B$, we may fix $\rho > 0$ small enough to guarantee that $Q \cap \operatorname{cl} \Omega_\rho(z^i) = \emptyset$ and, for any choice of $x^i \in B \cap \operatorname{cl} \Omega_\rho(z^i) \equiv \mathcal{U}_i$ and $y \in Q$, both $n(y)\cdot(y-x^1)$ and $n(y)\cdot(y-x^2)$ are non-zero and of the same sign, so that

$$\left|\frac{n(y)\cdot(y-x^1)}{|y-x^1|^m} - \frac{n(y)\cdot(y-x^2)}{|y-x^2|^m}\right| < \frac{|n(y)\cdot(y-x^1)|}{|y-x^1|^m} + \frac{|n(y)\cdot(y-x^2)|}{|y-x^2|^m} .$$

Since both functions occurring in this inequality are continuous on the compact set $Q \times \mathcal{U}_1 \times \mathcal{U}_2$ $(y \in Q,\ x^i \in \mathcal{U}_i)$, there is an $\varepsilon > 0$ such that

$$\left|\frac{n(y)\cdot(y-x^1)}{|y-x^1|^m} - \frac{n(y)\cdot(y-x^2)}{|y-x^2|^m}\right| + \varepsilon \leqq \frac{|n(y)\cdot(y-x^1)|}{|y-x^1|^m} + \frac{|n(y)\cdot(y-x^2)|}{|y-x^2|^m} .$$

Hence we conclude by (2), (3) (cf. also 2.15, 2.8, 2.9) that

$$\|\tau_{x^1} - \tau_{x^2}\| \leqq |2d_G(x^1)-1| + |2d_G(x^2)-1| +$$

$$+ |\lambda^G_{x^1} - \lambda^G_{x^2}|(B \setminus \{x^1,x^2\}) \leqq |2d_G(x^1)-1| + |2d_G(x^2)-1| -$$

$$- \frac{2\varepsilon}{A}\mathcal{H}_{m-1}(Q) + |\lambda^G_{x^1}|(B \setminus \{x^1\}) + |\lambda^G_{x^2}|(B \setminus \{x^2\}) =$$

$$= \|\tau_{x^1}\| + \|\tau_{x^2}\| - \frac{2\varepsilon}{A}\mathcal{H}_{m-1}(Q) = 2\left(1 - \frac{\varepsilon}{A}\mathcal{H}_{m-1}(Q)\right)$$

whenever $x^i \in \mathcal{U}_i$ and the proof is complete.

3.4. Corollary. If $P \subset B$ is a compact set such that

$$z^1, z^2 \in P \Longrightarrow Q_{z^1} \cap Q_{z^2} \neq C \quad (= \mathrm{cl}\, C),$$

then

$$\sup\{\mathrm{osc}\, Tf(P);\ f \in \mathcal{C}(B),\ \mathrm{osc}\, f \leqq 1\} < 1 .$$

Proof. According to 3.3 we may associate with any couple $z^1, z^2 \in P$ two sets $V(z^1)$, $V(z^2)$ open in B and a constant $q\,(V(z^1), V(z^2)) \in\,]0,1[$ such that $z^i \in V(z^i)$ and the following implication holds:

$$x^i \in V(z^i) \quad (i = 1,2) \Longrightarrow \|\tau_{x^1} - \tau_{x^2}\| \leq 2q(V(z^1),V(z^2)) .$$

In view of the compactness of $P \times P$ there exists a finite subcover $V_1^1 \times V_2^1, \dots, V_1^p \times V_2^p$ of $P \times P$ in the cover $\{V(z^1) \times V(z^2)\}_{(z^1,z^2) \in P \times P}$. Put $q = \max_{1 \leq j \leq p} q(V_1^j, V_2^j) \in$ $\in \,]0,1[$ and consider an arbitrary $f \in \mathcal{C}(B)$ with $\operatorname{osc} f \leq$ ≤ 1. Next choose $k \in R^1$ such that $|f-k| \leq \frac{1}{2} \operatorname{osc} f \leq \frac{1}{2}$ and fix $x^1 \in P$ with $Tf(x^1) = \max Tf(P)$ and $x^2 \in P$ with $Tf(x^2) = \min f(P)$. Clearly, $(x^1,x^2) \in V_1^j \times V_2^j$ for suitable j and, consequently,

$$\operatorname{osc} Tf(P) = \int_B f d(\tau_{x^1} - \tau_{x^2}) = \int_B (f-k) d(\tau_{x^1} - \tau_{x^2}) \leq$$

$$\leq \frac{1}{2} \|\tau_{x^1} - \tau_{x^2}\| \leq q .$$

3.5. Theorem. The implication (10) holds with a $q \in$ $\in [0,1[$ if and only if

$$Q_{z^1} \cap Q_{z^2} \neq C$$

for every couple of points $z^1, z^2 \in B$.

Proof. Letting $P = B$ in 3.4 we see that the condition is sufficient. Suppose now that there are points $z^1, z^2 \in B$ with

$$Q_{z^1} \cap Q_{z^2} = C ;$$

C being bounded we have necessarily $z^1 \neq z^2$. Next define g on B as follows:

$g(z^1) = 1$, $g = 1$ on $(B \setminus \{z^2\}) \cap \operatorname{int} Q_{z^1}$, $g = 0$ elsewhere on B . If $y \in B \setminus \{z^2\}$ and $g(y) = 0$, then necessarily either $n(y) = 0$ or else $n(y)\cdot(y-z^1) = 0$. Hence we conclude by (3) that $g = 1$ almost everywhere with respect to τ_{z^1} , so that

$$\int_B g d\,\tau_{z^1} = \tau_{z^1}(B) = 1 .$$

Observe that $n(y)\cdot(y-z^2) = 0$ for $y \in \partial Q_{z^2} \supset B \cap \operatorname{int} Q_{z^1}$, so that $g = 0$ almost everywhere with respect to τ_{z^2} (see (3)). Hence

$$\int_B g d\,\tau_{z^2} = 0 .$$

It is easy to see that there is a sequence of functions $f_n \in \mathcal{E}(B)$ with $0 \leqq f_n \leqq 1$ such that

$$\lim_{n\to\infty} f_n = g$$

pointwise on B . Validity of (10) would imply

$$|Tf_n(z^1) - Tf_n(z^2)| \leqq q \operatorname{osc} f_n \leqq q$$

for all n , which together with

$$\lim_{n\to\infty} Tf_n(z^i) = \lim_{n\to\infty} \int_B f_n d\tau_{z^i} = \int_B g d\tau_{z^i} \quad (i = 1,2)$$

yields

$$1 = \left|\int_B g d\tau_{z^1} - \int_B g d\tau_{z^2}\right| \leqq q .$$

We see that (10) is impossible with a $q < 1$ and the proof

is complete.

3.6. Remark. Theorem 3.5 shows that the Neumann operator $T = T^G$, considered on $\mathcal{C}(\partial G)/_{\mathcal{K}(\partial G)}$, is not contractive for simple convex bodies $C = R^m \setminus G$ resulting as an intersection of two cones with vertices in $B = \partial C$; in R^2 such exceptional convex bodies are triangles and quadrangles. Fortunately, the second iterate $T^2 \ldots = T(T\ldots)$ of the Neumann operator behaves well also for these exceptional bodies, as shown in Theorem 3.8 below. For its proof we shall need the following

3.7. Lemma. For every couple of points $z^1, z^2 \in B$ there exist constants $q \in [0,1[$ and $\rho > 0$ such that

$$(11) \qquad |T^2 f(x^1) - T^2 f(x^2)| \leqq q \operatorname{osc} f$$

whenever $f \in \mathcal{C}(B)$ and $|x^i - z^i| \leqq \rho$, $x^i \in B$ $(i = 1,2)$.

Proof. Since C is bounded, we have $B \cap \operatorname{int} Q_{z^i} \neq \emptyset$. In view of the assumption $B = \partial_e G$ we conclude from the isoperimetric lemma that a suitable closed ball contained in $\operatorname{int} Q_{z^i}$ meets $\hat{B}$ in a set of positive $\mathcal{H}_{m-1}$-measure (cf. 2.14). Employing Luzin's theorem we may thus fix a compact $Q^i \subset \hat{B} \cap \operatorname{int} Q_{z^i}$ with $\mathcal{H}_{m-1}(Q^i) > 0$ such that the vector--valued function

$$n : y \longmapsto n^G(y)$$

is continuous on Q^i. Since $|n(y)\cdot(y-z^i)| > 0$ for $y \in Q^i$, we may fix $\varrho_i > 0$ with $Q^i \cap \mathrm{cl}\, \Omega_{\varrho_i}(z^i) = \emptyset$ and $\varepsilon_i > 0$ such that

(12) $(y \in Q^i,\ x^i \in B,\ |x^i - z^i| \leqq \varrho_i) \Longrightarrow$

$$\Longrightarrow |n(y)\cdot(y-x^i)| / |y-x^i|^m \geqq A\,\varepsilon_i .$$

Put $P = Q^1 \cup Q^2$. Note that $Q_y = Q_y(C)$ is a half-space for any $y \in P$. C being bounded we have thus $Q_{y^1} \cap Q_{y^2} \neq C$ for any couple of points $y^1, y^2 \in P$. Using corollary 3.4 we get

(13) $\sup\{\mathrm{osc}\ Tf(P);\ f \in \mathcal{C}(B),\ \mathrm{osc}\ f \leqq 1\} = q_0 < 1$.

Consider now an arbitrary $f \in \mathcal{C}(B)$ with $f(B) = [0,1]$. We know that T is a positive operator on $\mathcal{C}(B)$ preserving constants, so that on B

$$0 \leqq Tf \leqq 1 .$$

Put $\beta = \min\{2\varepsilon_i \mathcal{H}_{m-1}(Q^i);\ i = 1,2\}$, $\varrho = \min(\varrho_1, \varrho_2)$ and consider arbitrary points $x^i \in B$ with $|x^i - z^i| \leqq \varrho$ $(i = 1,2)$. Since $\tau_{x^i}(Q^i) \geqq 2\varepsilon_i \mathcal{H}_{m-1}(Q^i) \geqq \beta$ by (3) and (12), we get for

(14) $r_f = \max Tf(P)$, $s_f = \min Tf(P)$

the estimates

$$T^2 f(x^i) = \Big(\int_{B \setminus Q^i} + \int_{Q^i}\Big) Tf\, d\tau_{x^i} \leqq \tau_{x^i}(B \setminus Q^i) + r_f \tau_{x^i}(Q^i) =$$

$$= \tau_{x^i}(B) - (1-r_f)\tau_{x^i}(Q^i) \leqq 1 - (1-r_f)\beta ,$$

$$T^2f(x^i) \geq \int_{Q^i} Tf\,d\tau_{x^i} \geq s_f\beta \;.$$

Employing (13), (14) we obtain $r_f - s_f \leq q_o$ and writing

$$q = 1 - (1-q_o)\beta$$

we get

$$|T^2f(x^1) - T^2f(x^2)| \leq 1 - [1-(r_f-s_f)]\beta \leq 1 - (1-q_o)\beta = q \;,$$

so that (11) is established for any $f \in \mathcal{E}(B)$ with $f(B) = [0,1]$. The rest is obvious.

<u>3.8.</u> <u>Theorem.</u> If $C = R^m \setminus G$ is an arbitrary convex body and $B = \partial C$, then the second iterate of the corresponding Neumann operator $T^G \equiv T$ is contractive on $\mathcal{E}(B)/_{\mathcal{H}(B)}$ in the sense that for suitable $q \in [0,1[$ the following implication holds:

(15) $f \in \mathcal{E}(B) \Longrightarrow \operatorname{osc} T^2f \leq q^2 \operatorname{osc} f$.

<u>Proof.</u> This follows at once from lemma 3.7 by a compactness argument similar to that used in the proof of 3.4.

<u>3.9.</u> <u>Theorem (on the interior Dirichlet problem).</u> Let C, B, T, G have the same meaning as in 3.8. If $h \in \mathcal{E}(B)$, then the series

(16) $$h + \sum_{n=1}^{\infty} (T^{2n}h - T^{2n-1}h)$$

is uniformly convergent on B and for suitable constant $c(h) \equiv c$

$$(17) \qquad \lim_{k\to\infty} T^k h = c\,\underline{1}$$

uniformly on B ; if $s(h) \equiv s$ denotes the sum of (16), then

$$(I + T)s = h + c\,\underline{1}\,.$$

Consequently, if $g \in \mathcal{E}(B)$ is arbitrarily prescribed and we put

$$h = 2g, \quad f = s(h) - \frac{1}{2}c(h) = 2g + \sum_{n=1}^{\infty}\left[T^{2n}(2g) - T^{2n-1}(2g)\right] - \lim_{k\to\infty} T^k g\,,$$

then for any $y \in B$

$$W^G f(y) = \lim_{\substack{x\to y\\ x\in \operatorname{int} C}} W^G f(x) = g(y)\,,$$

so that $W^G f$ represents a solution of the Dirichlet problem for $\operatorname{int} C$ and the boundary condition g .

<u>Proof.</u> Since T is a positive operator preserving constant functions we have for any $f \in \mathcal{E}(B)$ and any natural number k

$$T^k f(B) \subset T^{k-1} f(B)\,.$$

If we choose $x_M \in B$ so that

$$T^{k-1} f(x_M) = \max T^{k-1} f(B)$$

and $x_m \in B$ so that

$$T^{k-1} f(x_m) = \min T^{k-1} f(B)\,,$$

then

$$T^k f(x_M) - T^{k-1} f(x_M) \leqq 0\,, \quad T^k f(x_m) - T^{k-1} f(x_m) \geqq 0\,,$$

whence

$$0 \in (T^k f - T^{k-1} f)(B) .$$

Noting that

$$(g \in \mathcal{C}(B), \quad 0 \in g(B)) \Rightarrow \|g\| \leqq \operatorname{osc} g$$

we conclude from 3.8 that

$$\|T^{2n}h - T^{2n-1}h\| \leqq q^{2(n-1)} \operatorname{osc}(T^2 h - Th)$$

so that the series (16) is uniformly convergent. By 3.8 and 3.1 we get $\lim_{k \to \infty} \operatorname{osc} T^k f = 0$, so that the sequence $\{T^k f\}_{k=1}^{\infty}$ converges uniformly to a constant function. Writing

$$h_n = h + \sum_{k=1}^{n} (T^{2k}h - T^{2k-1}h)$$

we have

$$(I + T)h_n = h + T^{2n+1}h$$

and making $n \to \infty$ we arrive at

$$(I + T)s(h) = h + c(h)\,\underline{1} ,$$

whence

$$(I + T)(s - \tfrac{1}{2} c\,\underline{1}) = h .$$

It remains to recall that, by 2.19,

$$\lim_{\substack{x \to y \\ x \in \operatorname{int} C}} W^G f(x) = \tfrac{1}{2} (I + T)f(y)$$

for every $y \in B$.

3.10. Proposition. Let us keep the notation of 3.9. Let T' denote the dual operator (acting on $\mathcal{C}'(B)$) to T . Then there is a uniquely determined $\rho \in \mathcal{C}'(B)$ with

(18) $\quad T'\rho = \rho \, , \quad \rho(B) = 1 \, .$

This measure ρ is non-negative, its potential $\mathcal{U}\rho$ is continuous on R^m and constant on C (so that ρ represents the so-called equilibrium probability distribution for C) and for any $\mu \in \mathcal{C}'(B)$

(19) $$\lim_{n\to\infty} \|(T')^n\mu - \mu(B)\rho\| = 0 \, .$$

If $c \equiv c(h)$ is defined by (17), then

(20) $$h \in \mathcal{C}(B) \Rightarrow \int_B h\,d\rho = c(h) \, .$$

Proof. Since T is a positive operator preserving constant functions, the linear functional

$$h \mapsto c(h)$$

is positive on $\mathcal{C}(B)$ and assumes the value 1 at $\underline{1}$. Consequently, there is a uniquely determined $\rho \in \mathcal{C}'(B)$ satisfying (20); clearly, $\rho \geqq 0$ and $\rho(B) = 1$. For any $h \in \mathcal{C}(B)$, the quantity $\langle h, T'\rho\rangle = \langle Th, \rho\rangle = c(Th)$ represents the value attained by $\lim_{n\to\infty} T^{n+1}h = \lim_{n\to\infty} T^n h$ on B, so that $\langle h, T'\rho\rangle = \langle h,\rho\rangle$ and (18) is verified. Put

$$\mathcal{C}'_0(B) = \{\mu \in \mathcal{C}'(B);\ \mu(B) = 0\} \, .$$

If $\mu \in \mathcal{C}'_0(B)$, then $\langle \underline{1}, T'\mu\rangle = \langle T\,\underline{1}, \mu\rangle = 0$, so that $T'\mu \in \mathcal{C}'_0(B)$; we see that

(21) $$T'(\mathcal{C}'_0(B)) \subset \mathcal{C}'_0(B) \, .$$

Note that

(22) $\mu \in \mathcal{C}_0'(B) \Longrightarrow \|\mu\| = \sup\{\langle g,\mu\rangle \,;\, g \in \mathcal{C}(B),$
$\text{osc } g \leqq 2\}.$

Indeed, if $g \in \mathcal{C}(B)$ and $\text{osc } g \leqq 2$, then there is a $c \in R^1$ with $\|g - c\,\mathbb{1}\| \leqq 1$, so that

$$\langle g,\mu\rangle = \langle g - c\,\mathbb{1}, \mu\rangle$$

and

$$\sup\{\langle g,\mu\rangle \,;\, \text{osc } g \leqq 2\} \leqq \sup\{\langle f,\mu\rangle \,;\, f \in \mathcal{C}(B),\ \|f\| \leqq 1\} = \\ = \|\mu\| \,;$$

conversely, if $f \in \mathcal{C}(B)$ and $\|f\| \leqq 1$, then $\text{osc } f \leqq 2$ and $\langle f,\mu\rangle \leqq \sup\{\langle g,\mu\rangle \,;\, \text{osc } g \leqq 2\}$ which gives the opposite inequality and proves (22). Employing (22) and 3.8 we get for any $\mu \in \mathcal{C}_0'(B)$ $\|(T')^2\mu\| = \sup\{\langle g,(T')^2\mu\rangle \,;$ $\text{osc } g \leqq 2\} = \sup\{\langle T^2 g,\mu\rangle \,;\, \text{osc } g \leqq 2\} \leqq \sup\{\langle h,\mu\rangle \,;$ $\text{osc } h \leqq 2q^2\} = q^2\|\mu\|$, where $q \in [0,1[$. We have thus verified the implication

(23) $\mu \in \mathcal{C}_0'(B) \Longrightarrow \|(T')^2\mu\| \leqq q^2\|\mu\|$

for suitable $q \in [0,1[$. Clearly, $\|T'\| = \|T\| \leqq 1$ (cf. 3.1) and (23) yields

(24) $\nu \in \mathcal{C}_0'(B) \Longrightarrow \lim\limits_{n\to\infty} \|(T')^n \nu\| = 0$.

If $\mu \in \mathcal{C}'(B)$, then

$$(T')^n\mu - \mu(B)\rho = (T')^n[\mu - \mu(B)\rho]$$

with $\mu - \mu(B)\rho = \nu \in \mathcal{C}_0'(B)$, so that (24) implies (19). If, besides that, $\mu(B) = 1$ and $T'\mu = \mu$, then $(T')^n\mu = \mu$ and (19) gives $\|\mu - \rho\| = 0$, so that the conditions (18) determine ρ uniquely. Let us now define $\nu \in \mathcal{C}'(B)$ by

$$d\nu(y) = c|n(y)|\cdot d\mathcal{H}_{m-1}(y)\ , \quad y \in B\ ,$$

where the vector-valued function $n(y)$ has the meaning described in 2.14 and the constant $c > 0$ is chosen in such a way that $\nu(B) = 1$. Employing 2.17 and 2.18 we conclude that the potential $\mathcal{U}\nu$ is continuous on R^m. Let now $\mathcal{C}_c'(B)$ have the meaning described in 2.14 and, for $\mu \in \mathcal{C}_c'(B)$, define $\mathcal{U}_o\mu$ ($\in \mathcal{C}(B)$) as in 2.23. Employing Plemelj's theorem 2.23 and the equality $T = 2W_B - I$ we obtain that

$$T'(\mathcal{C}_c'(B)) \subset \mathcal{C}_c'(B)$$

and

$$\mu \in \mathcal{C}_c'(B) \Longrightarrow \mathcal{U}_o(T'\mu) = T(\mathcal{U}_o\mu) \quad \text{on} \quad B\ .$$

Hence we conclude easily that, for every positive integer n, $(T')^n\nu \in \mathcal{C}_c'(B)$ has potential $\mathcal{U}[(T')^n\nu]$ whose values on $R^m \setminus B$ extend continuously to the values given on B by $T^n(\mathcal{U}_o\nu)$. Note that, for suitable constant $t \in R^1$, $\lim_{n\to\infty} T^n(\mathcal{U}_o\nu) = t\,\underline{1}$ uniformly on B (cf. 3.9). Fix now a large ball $\Omega = \Omega_R(0)$ such that $\mathrm{cl}\, C \subset \Omega$. Since

$$\lim_{n\to\infty} \|(T')^n\nu - \rho\| = 0\ ,$$

we conclude that $\mathcal{U}[(T')^n\nu]$ converge to $\mathcal{U}\rho$ locally uniformly on $R^m \setminus B$ and uniformly on $\partial\Omega$ as $n \to \infty$. Consider now the sequence of harmonic functions $\mathcal{U}[(T')^n\nu]$ on $V = \Omega \setminus B$. They all extend continuously from V to $\mathrm{cl}\, V$ and the extended functions converge (as $n \to \infty$) on ∂V uniformly to the continuous function equal to $t\cdot\underline{1}$ on

B and to $\mathcal{U}\rho$ on ∂V. This implies that the limit function $\mathcal{U}\rho$ on V extends continuously to the values given by $t\cdot\underline{1}$ on B and $\mathcal{U}\rho$ on $\partial\Omega$. By the maximum-minimum principle for harmonic functions this means that $\mathcal{U}\rho = t$ on int C. In view of lower-semicontinuity of $\mathcal{U}\rho$ we get for any $y \in B$ the inequality $\mathcal{U}\rho(y) \leqq t$. On the other hand, noting that $\mathcal{H}_m(B) = 0$ and $\lim\limits_{\substack{x\to y\\ x\in R^m\setminus B}} \mathcal{U}\rho(x) = t$, we conclude from superharmonicity of $\mathcal{U}\rho$ that

$$\mathcal{U}\rho(y) \geqq \lim_{r\to 0+} \frac{\int_{\Omega_r(y)} \mathcal{U}\rho(x)\,dx}{\mathcal{H}_m(\Omega_r(y))} \geqq t\,,$$

so that $\mathcal{U}\rho(y) = t$. Consequently, $\mathcal{U}\rho = t$ on cl C and $\mathcal{U}\rho$ is continuous on R^m.

<u>3.11.</u> <u>Theorem (on the exterior Neumann problem).</u> Let $C \subset R^m$ be a convex body, $G = R^m \setminus C$, $B = \partial C$, $T = T^G$. Then for each $\mu \in \mathcal{E}'(B)$ the series

$$(25)\qquad \mu + \sum_{n=1}^{\infty}\left[(T')^{2n}\mu - (T')^{2n-1}\mu\right]$$

converges with respect to the norm $\|\ldots\|$ in $\mathcal{E}'(B)$ to a certain $\sigma(\mu) \in \mathcal{E}'(B)$. Besides that, the relation (19) holds (where ρ has the meaning described in 3.10) and

$$2N^G\mathcal{U}\sigma(\mu) \equiv (I + T)'\sigma(\mu) = \mu + \mu(B)\rho\ .$$

Consequently, if $\nu \in \mathcal{E}'(B)$ is arbitrarily prescribed and we put

$$\mu = \sigma(2\nu) - \nu(B)\rho ,$$

then

$$N^G \mathcal{U}\mu = \nu .$$

Proof. If $\mu \in \mathcal{C}'(B)$ and k is a positive integer, then $\langle 1, (T')^{2k}\mu - (T')^{2k-1}\mu \rangle = \langle T^{2k} 1, \mu \rangle - \langle T^{2k-1} 1, \mu \rangle = \mu(B) - \mu(B) = 0$, so that $(T')^{2k}\mu - (T')^{2k-1}\mu \in \mathcal{C}'_o(B)$. Employing (23) we get

$$\|(T')^{2k}\mu - (T')^{2k-1}\mu\| \leq q^{2(k-1)} \|(T')^2\mu - T'\mu\|$$

for suitable $q \in [0, 1[$ and the convergence of (25) to a certain $\sigma(\mu) \in \mathcal{C}'(B)$ follows. If

$$\mu_n = \mu + \sum_{k=1}^{n} [(T')^{2k}\mu - (T')^{2k-1}\mu] ,$$

then

$$(I + T)' \mu_n = \mu + (T')^{2n+1}\mu .$$

Making $n \to \infty$ and using (19) we obtain

$$(I + T)' \sigma(\mu) = \mu + \mu(B)\rho$$

or, which is the same,

$$(I + T)' (\sigma(\mu) - \tfrac{1}{2}\mu(B)\rho) = \mu .$$

It remains to recall that $N^G \mathcal{U} = \frac{1}{2}(I + T)'$.

3.12. Remark. Let Q_1, Q_2 be disjoint open sets in R^m and suppose that $B = R^m \setminus Q_1 \setminus Q_2$ is compact and satisfies the assumptions

$$\mathcal{H}_m(B) = 0 , \quad B \subset \text{cl } Q_1 \cap \text{cl } Q_2$$

(so that B is the common boundary of both Q_1 and Q_2).

If $V^{Q_1} < +\infty$ (cf. (32) in §2), then

$$\mu \in \mathcal{E}'(B) \Longrightarrow N^{Q_1}\mathcal{U}\mu + N^{Q_2}\mathcal{U}\mu = \mu .$$

<u>Proof.</u> If $\mu \in \mathcal{E}'(B)$ and $\varphi \in \mathcal{D}$, then 1.5 yields

$$\langle \varphi, N^{Q_1}\mathcal{U}\mu + N^{Q_2}\mathcal{U}\mu \rangle = \int_B (\langle \varphi, N^{Q_1}\mathcal{U}\delta_y \rangle +$$

$$+ \langle \varphi, N^{Q_2}\mathcal{U}\delta_y \rangle) d\mu(y) .$$

For any fixed $y \in B$ we get according to the example following 1.4

$$\langle \varphi, N^{Q_1}\mathcal{U}\delta_y \rangle + \langle \varphi, N^{Q_2}\mathcal{U}\delta_y \rangle =$$

$$= \int_{Q_1 \cup Q_2} \operatorname{grad} \varphi(x) \cdot \operatorname{grad} h_y(x) dx =$$

$$= \int_{R^m \setminus \{y\}} \operatorname{grad} \varphi(x) \cdot \operatorname{grad} h_y(x) dx = \varphi(y) ,$$

so that

$$\langle \varphi, N^{Q_1}\mathcal{U}\mu + N^{Q_2}\mathcal{U}\mu \rangle = \langle \varphi, \mu \rangle .$$

<u>3.13.</u> <u>Theorem (on the interior Neumann problem).</u> Let $C \subset R^m$ be a convex body, $B = \partial C$, $Q = \operatorname{int} C$ $(= C \setminus B)$. Given $\mu \in \mathcal{E}'(B)$, then a necessary and sufficient condition for the existence of a $\nu \in \mathcal{E}'(B)$ with

(26) $$N^Q \mathcal{U} \nu = \mu$$

reads as follows:

(27) $$\mu(B) = 0 \quad (\Longleftrightarrow \mu \in \mathcal{C}_0'(B)) .$$

If (27) holds and $T = T^G$, where $G = R^m \setminus C$, then a solution of (26) is given by the series

(28) $$(2\mu) + \sum_{n=1}^{\infty} (T')^n (2\mu)$$

which is convergent with respect to the norm $\|\ldots\|$ in $\mathcal{C}'(B)$.

The solution of (26) is not uniquely determined in $\mathcal{C}'(B)$: any other solution differs from that given by (28) by a constant multiple of the equilibrium distribution ρ (cf. 3.10); (28) gives the unique solution of (26) in $\mathcal{C}_0'(B)$.

Proof. We have observed in 3.12 that $N^Q \mathcal{U} = I' - N^G \mathcal{U}$. Since $N^G \mathcal{U} = (W_B^G)'$ (cf. 2.20), we conclude from (51) in §2 that

$$N^Q \mathcal{U} = \frac{1}{2} (I - T)' ,$$

so that (26) is equivalent with

(29) $$(I - T)' \nu = 2\mu .$$

If (29) holds, then $\mu(B) = \frac{1}{2} \langle \underline{1}, (I-T)' \nu \rangle =$
$= \frac{1}{2} \langle \underline{1} - T \underline{1}, \nu \rangle = 0$. Conversely, if (27) holds, then $\mu_1 = 2\mu \in \mathcal{C}_0'(B)$ and (23) together with 3.1 yields the estimates

$$\|(T')^{2n+1} \mu_1\| \leqq \|(T')^{2n} \mu_1\| \leqq q^{2n} \|\mu_1\|$$

for suitable $q \in [0,1[$, which guarantee the convergence of the series (28) in $\mathscr{C}'(B)$; easy reasoning (analogous to that used in the proof of 3.11) shows that the corresponding sum provides a $\nu \in \mathscr{C}'_0(B)$ satisfying (29). It remains to observe that, in view of 3.10,

$$(I - T)'\nu = 0 \quad (\in \mathscr{C}'(B)) \Longrightarrow \nu = \nu(B)\rho \ .$$

__3.14.__ __Theorem.__ Let C, G, B, T have the same meaning as in 3.8 and denote by $\rho \in \mathscr{C}'(B)$ the equilibrium distribution for C defined by (18). Given $g \in \mathscr{C}(B)$, then a necessary and sufficient condition for the existence of an $f \in \mathscr{C}(B)$ satisfying

$$\lim_{\substack{x \to y \\ x \in G}} W^G f(x) = g(y) \quad \text{for all} \quad y \in B \tag{30}$$

reads as follows:

$$\int_B g \, d\rho = 0 \ . \tag{31}$$

If (31) holds and $g_1 = -2g$, then the series

$$g_1 + \sum_{n=1}^{\infty} T^n g_1 \tag{32}$$

converges uniformly on B and its sum provides an $f \in \mathscr{C}(B)$ with (30). The solution f of (30) is not uniquely determined in $\mathscr{C}(B)$: any other solution differs from that given by (32) by a constant function in $\mathscr{K}(B)$; (32) gives the unique solution of (30) in the subspace

$$\{f \in \mathscr{C}(B);\ \langle f, \rho \rangle = 0\}. \tag{33}$$

Proof. We know from 2.19 that, for $f \in \mathcal{C}(B)$ and $y \in B$,

$$\lim_{\substack{x \to y \\ x \in G}} W^G f(x) = W^G_B f(y) - f(y) = \frac{1}{2}(T-I)f(y)$$

(cf. also (51) in §2). This means that (30) is equivalent with

(34) $$(I - T)f = -2g \equiv g_1 .$$

If (34) holds, then necessarily $-2\langle g, \rho\rangle = \langle f, \rho\rangle - \langle Tf, \rho\rangle = \langle f, \rho\rangle - \langle f, T'\rho\rangle = 0$. Conversely, if (31) holds and $g_1 = -2g$, then for each positive integer n

$$\langle T^n g_1, \rho\rangle = \langle g_1, (T')^n \rho\rangle = \langle g_1, \rho\rangle = 0 ,$$

so that $0 \in T^n g_1(B)$ and

$$\| T^n g_1 \| \leqq \operatorname{osc} T^n g_1 .$$

This together with (15) and 3.1 provides the estimates

$$\| T^{2n+1} g_1 \| \leqq \| T^{2n} g_1 \| \leqq q^{2n} \| g_1 \|$$

with a suitable $q \in [0,1[$ guaranteeing uniform convergence of the series (32). It is easily seen that the sum f of (32) belongs to the subspace (33) and satisfies (34). We know from 2.8 (cf. (11)) that $W^G f_o(z) = 0$ for $z \in G$ and $f_o \in \mathcal{K}(B)$ (cf. 3.2). Conversely, let $h \in \mathcal{C}(B)$ satisfy

$$(I - T)h = 0 \quad (\in \mathcal{C}(B)) .$$

Then, for each positive integer k,

$$(I - T^k)h = (I + T + \ldots + T^{k-1})(I - T)h = 0$$

whence we conclude by (17) that

$$h = \lim_{k \to \infty} T^k h \in \mathcal{H}(B) .$$

We see that

$$\mathcal{H}(B) = \{f_o \in \mathcal{E}(B);\ (I-T)f_o = 0\} .$$

The rest is obvious.

<u>Remark.</u> We have observed in 3.14 that the solution of the Dirichlet problem for G cannot, in general, be represented in the form of a double layer potential $W^G f$ (this being possible only if the prescribed boundary condition g satisfies (31)). Nevertheless, if $m > 2$, one can always represent this solution as a combination of a double layer potential and a fundamental harmonic function with pole off $\mathrm{cl}\, G$ as shown by the following

<u>3.15. Corollary (on the exterior Dirichlet problem).</u> Let $C \subset R^m$ be a convex body, $G = R^m \setminus C$, $T = T^G$ and denote by ρ the equilibrium distribution for C characterized by (18). Fix $z \in \mathrm{int}\, C$ and put $c = \mathcal{U}\rho(z)$ (so that c is the value of $\mathcal{U}\rho$ on the whole C). Assume $c \neq 0$ (which is always the case if $m > 2$). Given $g \in \mathcal{E}(B)$, we put $c_g = \frac{1}{c} \int_B g\, d\rho$ and define $g_1 = -2(g - c_g h_z)$ on B . Then the series (32) converges uniformly on B to an $f \in \mathcal{E}(B)$ and defining

$$F(x) = W^G f(x) + c_g h_z(x) , \quad x \in G ,$$

we get a harmonic function on G with

$$\lim_{\substack{x \to y \\ x \in G}} F(x) = g(y) \quad \text{for all} \quad y \in B .$$

Proof. The function $\tilde{g} = g - c_g h_z$ satisfies

$$\langle \tilde{g}, \rho \rangle = \langle g, \rho \rangle - c_g \mathcal{U}\rho(z) = 0 ,$$

so that we may apply 3.14 with g replaced by $\tilde{g}$. The rest is obvious.

3.16. Remark. If $C = R^m \setminus G$ is a convex body, then the m-dimensional density $d_G(z)$ is defined for all points in $B = \partial G$ (and, consequently, for all $z \in R^m$) and it is easily seen that, for suitable $\gamma > 0$,

$$y \in B \Longrightarrow \frac{1}{2} \leqq d_G(y) \leqq 1 - \gamma .$$

Clearly, $v^G(y) \leqq 1$ for any $y \in B$ (cf. 1.11, 1.12). The reduced boundary $\hat{B}$ of G (as defined in 2.14) now coincides with $\{y \in R^m;\ d_G(y) = \frac{1}{2}\}$ which is precisely the set of those $y \in B$ at which there exists a unique supporting hyperplane for C . Employing the isoperimetric inequality (cf. 2.14) we observe that for any $y \in B$ and suitable $\alpha > 0$, $\varepsilon > 0$

$$0 < r < \varepsilon \Longrightarrow \mathcal{H}_{m-1}(\Omega_r(y) \cap \hat{B}) \geqq \alpha r^{m-1} .$$

Hence it follows (cf. §3 in [Fe3]) that

(35) $$\mathcal{H}_{m-1}(B \setminus \hat{B}) = 0 .$$

These facts will be useful below.

3.17. Proposition. Let us keep the notation of 3.8 and let $\mathcal{C}_a'(B)$ stand for the subspace of all $\mu \in \mathcal{C}'(B)$ that are absolutely continuous with respect to $\mathcal{H}_{m-1}$. Then

(36) $$T'(\mathcal{C}_a'(B)) \subset \mathcal{C}_a'(B) .$$

Proof. Let $\mu \in \mathcal{C}'_a(B)$ and consider $f \in \mathcal{C}(B)$. Employing (3) and noting that, in view of (35),

$$|\mu|(\{z \in B;\ d_G(z) \neq \tfrac{1}{2}\}) = 0 ,$$

we get

$$\langle f, T'\mu \rangle = \langle Tf, \mu \rangle = \int_B \left(\int_B f\, d\tau_z \right) d\mu(z) =$$

$$= \int_B f(z) \left[2d_G(z) - 1 \right] d\mu(z) -$$

$$- 2 \int_B \left(\int_B f(y) n^G(y) \cdot \operatorname{grad} h_z(y)\, d\mathcal{H}_{m-1}(y) \right) d\mu(z) =$$

$$= - 2 \int_B \left(\int_B f(y) n^G(y) \cdot \operatorname{grad} h_z(y)\, d\mathcal{H}_{m-1}(y) \right) d\mu(z) .$$

According to 2.15 we have

$$\iint_{B \times B} |f(y)| \cdot |n^G(y) \cdot \operatorname{grad} h_z(y)|\, d\mathcal{H}_{m-1}(y)\, d|\mu|(z) \leqq$$

$$\leqq \|f\| \cdot \|\mu\| \cdot \sup_{z \in B} v^G(z) \leqq \|f\| \cdot \|\mu\| < +\infty ,$$

so that we are justified to apply Fubini's theorem concluding that

$$\langle f, T'\mu \rangle = - 2 \int_B f(y) \left(\int_B n^G(y) \cdot \operatorname{grad} h_z(y)\, d\mu(z) \right) d\mathcal{H}_{m-1}(y) .$$

We see that $T'\mu \in \mathcal{C}'_a(B)$ and its density with respect to $\mathcal{H}_{m-1}$ is given by the function

$$y \mapsto - 2 \int_B n^G(y) \cdot \operatorname{grad} h_z(y) d\mu(z)$$

which is defined a.e. and integrable with respect to $\mathcal{H}_{m-1}$ over B .

3.18. Corollary. If ϱ denotes the equilibrium distribution for the convex body C as defined by (18) in 3.10, then $\varrho \in \mathcal{C}'_a(B)$.

<u>Proof.</u> Let μ denote the constant multiple of $\mathcal{H}_{m-1}$ on B normalized in such a way that $\mu(B) = 1$. Then $\mu \in \mathcal{C}'_a(B)$ and, by (19), $(T')^n \mu \to \varrho$ with respect to the norm $\| \ldots \|$ in $\mathcal{C}'(B)$. Since all the measures $(T')^n \mu$ are absolutely continuous with respect to $\mathcal{H}_{m-1}$, the singular part in the Lebesgue decomposition of ϱ with respect to $\mathcal{H}_{m-1}$ on B must vanish, so that $\varrho \in \mathcal{C}'_a(B)$.

A similar reasoning in combination with theorems 3.11, 3.13 and 3.18 yields the following

3.19. Corollary. Let Q denote one of the sets $\operatorname{int} C$, $G = R^m \smallsetminus C$, where $C \subset R^m$ is a convex body, $B = \partial C$. If

$$N^Q \mathcal{U} \mu = \nu$$

and $\nu \in \mathcal{C}'_a(B)$, then also $\mu \in \mathcal{C}'_a(B)$.

3.20. Remark. We observed in the beginning of this paragraph that the Dirichlet problem for $\operatorname{int} C$ and the Neumann problem for $G = R^m \smallsetminus C$ reduce to a pair of dual equations on $B = \partial G$ of the form

(37) $(I + T)f = g_1$ (with prescribed $g_1 \in \mathcal{C}(B)$ and unknown $f \in \mathcal{C}(B)$),

(37′) $(I + T)'\nu = \mu_1$ (with prescribed $\mu_1 \in \mathcal{C}'(B)$ and unknown $\nu \in \mathcal{C}'(B)$)

and later we were able to prove that the corresponding operators $I + T$ and $(I + T)' = I' + T'$ are invertible on $\mathcal{C}(B)$ and $\mathcal{C}'(B)$, respectively. (In 3.9 we obtained a concrete expression for the inverse

$$(I + T)^{-1} = I + \sum_{n=1}^{\infty} (T^{2n} - T^{2n-1}) - \frac{1}{2} P ,$$

where $P \ldots = \lim_{n\to\infty} T^n \ldots$ is the finite-dimensional operator on $\mathcal{C}(B)$ given by

$$P \ldots = \langle \ldots, \varrho \rangle \underline{1} .$$

In 3.11 we showed that

$$(I' + T')^{-1} = I' + \sum_{n=1}^{\infty} \left[(T')^{2n} - (T')^{2n-1}\right] - \frac{1}{2} P' ,$$

where $P' \ldots = \lim_{n\to\infty} (T')^n \ldots$ is the finite dimensional operator on $\mathcal{C}'(B)$ given by

$$P' \ldots = \langle \underline{1}, \ldots \rangle \varrho .)$$

Similarly, the Dirichlet problem for G and the Neumann problem for $\operatorname{int} C$ led us to a pair of dual equations

(38) $(I - T)f = g_1$ ($g_1 \in \mathcal{C}(B)$ prescribed, $f \in \mathcal{C}(B)$ unknown),

(38′) $(I - T)'\nu = \mu_1$ ($\mu_1 \in \mathcal{C}'(B)$ prescribed, $\nu \in$ $\in \mathcal{C}'(B)$ unknown).

Now the corresponding homogeneous equations possess non-trivial solutions

$$\{f_o \in \mathcal{C}(B);\ (I-T)f_o = 0\} = \mathcal{K}\ ,$$

$$\{\nu_o \in \mathcal{C}'(B);\ (I-T)'\ \nu_o = 0\} = \{c\varrho\ ;\ c \in R^1\} \equiv \mathcal{N}\ ,$$

and I-T is invertible if restricted to the subspace

$$\mathcal{N}^{\perp} = \{f \in \mathcal{C}(B);\ \langle f,\varrho \rangle = 0\}$$

(where its inverse is given by the Neumann series $I + \sum_{n=1}^{\infty} T^n$), while $I' - T'$ is invertible if restricted to the subspace

$$\mathcal{K}^{\perp} \equiv \mathcal{C}_o'(B) = \{\nu_o \in \mathcal{C}'(B);\ \langle 1, \nu_o \rangle = 0\}$$

(the inverse being again given by the Neumann series $I' + \sum_{n=1}^{\infty} (T')^n$).

Our proofs of these results in this paragraph were based on contractivity of the second iterate of the Neumann operator established for convex C . However, contractivity of T^2 is not indispensable for tractability of the equations (37), (37′) and (38), (38′). The so-called Riesz-Schauder theory guarantees validity of Fredholm's theorems on solvability of such dual pairs of equations under the assumption that the operator T is compact. Consequently, it will be important for us to know under which general geometric conditions on the sets G, C can be asserted that the corresponding Neumann operator T is compact (or sufficiently closely approximable by compact operators). This will be the object of our investigation in the next paragraph.

§ 4

Fredholm radius of the Neumann operator

In this paragraph we always assume that $G \subset R^m$ and $C = R^m \setminus G$ are Borel sets with a common compact boundary B such that

$$(1) \quad (y \in B, \quad r > 0) \Longrightarrow \mathcal{H}_m(\Omega_r(y) \cap G) > 0 \text{ and } \mathcal{H}_m(\Omega_r(y) \cap C) > 0 .$$

We shall always assume that the quantity V^G defined by (32) in §2 is finite. According to 2.19, 2.20 this guarantees that, for each $f \in \mathcal{C}(B)$, the double layer potential $W^G f$ is defined and bounded on the whole R^m and the operator W_B^G (cf. (39) in §2) sending f into the restriction of $W^G f$ to B is bounded on $\mathcal{C}(B)$.

Let us recall that a linear operator V acting on $\mathcal{C}(B)$ is called compact if it maps each (norm-) bounded subset of $\mathcal{C}(B)$ onto a relatively compact set in $\mathcal{C}(B)$. By the Arzelà-Ascoli theorem (cf. chap. IV in [DS]), V is compact if and only if all the functions in

$$(2) \qquad \{Vg;\ g \in \mathcal{C}(B),\ \|g\| \leqq 1\}$$

are equicontinuous and uniformly bounded on B. We shall denote by $\mathcal{C}\!f$ the class of all compact linear operators acting on $\mathcal{C}(B)$; $\mathcal{C}\!f$ is a closed linear subspace in the space of all bounded linear operators on $\mathcal{C}(B)$ with the usual norm. If V is a bounded linear operator on $\mathcal{C}(B)$

and, for any $y \in B$, $\nu_y^V \in \mathcal{C}'(B)$ is defined by

(3) $\quad Vf(y) = \langle f, \nu_y^V \rangle \ , \quad f \in \mathcal{C}(B)$,

then Radon's results (compare chap. V in [RS]) assert that V is compact iff

$$\|\nu_y^V - \nu_z^V\| \to 0 \quad \text{as} \quad |y-z| \to 0+ \ , \quad y, z \in B \ .$$

Hence it follows easily that each $V \in \mathcal{V}$ can be arbitrarily closely approximated by finite dimensional operators. Indeed, given $\varepsilon > 0$, we can first fix a $\delta > 0$ such that

$$(y, z \in B \ , \ |y-z| < \delta) \Longrightarrow \|\nu_y^V - \nu_z^V\| < \varepsilon \ .$$

Further we choose non-negative functions $\varphi_k \in \mathcal{C}(B)$ $(k = 1,\dots,q)$ such that

$$\sum_k \varphi_k = \mathbb{1} \ , \quad \text{diam spt } \varphi_k < \delta$$

and fix $y_k \in \text{spt } \varphi_k$ $(k = 1,\dots,q)$. Defining the operator $\hat{V}$ by

$$\hat{V}f = \sum_{k=1}^{q} \langle f, \nu_{y_k}^V \rangle \varphi_k \ , \quad f \in \mathcal{C}(B) \ ,$$

we get for each $y \in B$ and any $f \in \mathcal{C}(B)$

$$|(V-\hat{V})f(y)| = \Big| \sum_{\substack{k \\ y \in \text{spt } \varphi_k}} \varphi_k(y) \int_B f d(\nu_y^V - \nu_{y_k}^V) \Big| \leqq \varepsilon \|f\| \ ,$$

so that $\|V - \hat{V}\| \leqq \varepsilon$.

For each bounded linear operator W acting on $\mathcal{C}(B)$

we denote by

(4) $$\omega W = \inf \{ \|W-V\| \; ; \; V \in \mathcal{C}\!\mathcal{f} \}$$

its distance from the subspace $\mathcal{C}\!\mathcal{f}$. The reciprocal of ωW (which is set to be $+\infty$ if $\omega W = 0$) is usually called the Fredholm radius of W. Our main objective in this paragraph is to evaluate the Fredholm radius of the Neumann operator $T^G = 2(W_B^G - \frac{1}{2} I)$ in geometric terms connected with G.

<u>4.1. Theorem.</u> Let

(5) $$V_o^G = \lim_{r \to 0+} \sup_{y \in B} v_r^G(y) ,$$

where $v_r^G(.)$ is defined in 1.11. Then for each $\alpha \in R^1$

(6) $$\omega(W_B^G - \alpha I) = |\tfrac{1}{2} - \alpha| + V_o^G =$$

$$= \lim_{r \to 0+} \sup_{y \in B} (|d_G(y) - \alpha| + v_r^G(y))$$

(here I is the identity operator on $\mathcal{C}(B)$ and $d_G(y)$ denotes the density of G at y as defined in (22) in §2).

<u>Proof.</u> Fix $R > 0$ and construct a function $\psi \equiv \psi^R$ on R^m such that the following conditions hold:

$$0 \leqq \psi \leqq 1 , \quad \psi(\Omega_R(0)) = \{1\} , \quad \psi(R^m \setminus \Omega_{2R}(0)) = \{0\} ,$$

$$x^1, x^2 \in R^m \Longrightarrow |\psi(x^1) - \psi(x^2)| \leqq |x^1 - x^2| .$$

Put for $y \in R^m$

$$\psi_y(x) = \psi(x-y) , \quad x \in R^m ,$$

and define the operator $V \equiv V^R$ on $\mathcal{C}(B)$ by

$$Vf(y) = \int_B f(1-\psi_y)\,d\,\lambda_y^G\ , \quad f \in \mathcal{C}(B)\ , \quad y \in B\ ,$$

where $\lambda_y^G \in \mathcal{C}'(B)$ is defined by 2.5. Since $1-\psi_y$ vanishes near y, we get by (12) in §2 that

(7) $|Vf(y)| \leqq \|f\| v^G(y) \leqq \|f\| \cdot V^G\ , \quad y \in B\ .$

We are going to prove that $V \in \mathcal{G}$. Consider an arbitrary $g \in \mathcal{C}(B)$ with $\|g\| \leqq 1$. We have then for any couple of points $y_1, y_2 \in B$

(8) $$Vg(y_1) - Vg(y_2) = \int_B g(\psi_{y_2} - \psi_{y_1})\,d\,\lambda_{y_1}^G + \int_B g(1-\psi_{y_2})\,d(\lambda_{y_1}^G - \lambda_{y_2}^G)\ .$$

It follows from the construction of ψ that

$$\|\psi_{y_2} - \psi_{y_1}\| \leqq |y_2 - y_1|\ ;$$

if $|y_1 - y_2| < R$ then, besides that, $(\psi_{y_2} - \psi_{y_1})(y_1) = 0$, so that we get by (12) in §2 the estimate

(9) $$\left|\int_B g(\psi_{y_2} - \psi_{y_1})\,d\,\lambda_{y_1}^G\right| \leqq |y_1 - y_2| \cdot V^G\ .$$

Employing 2.15 we obtain for $|y_1 - y_2| < R$

$$\int_B g(1-\psi_{y_2})\,d(\lambda_{y_1}^G - \lambda_{y_2}^G) =$$

$$= \int_{B \setminus \Omega_R(y_2)} g(1-\psi_{y_2}) n^G \cdot (\operatorname{grad} h_{y_2} - \operatorname{grad} h_{y_1}) \, d\mathcal{H}_{m-1} ,$$

where n^G is the vector-valued function defined in 2.14. It is easily seen that there is a constant $k \equiv k(R,m)$ (depending on R, m only) such that

$$(x \in B \setminus \Omega_R(y_2),\ |y_1-y_2| \leqq \tfrac{1}{2} R) \Longrightarrow$$

$$\Longrightarrow |\operatorname{grad} h_{y_2}(x) - \operatorname{grad} h_{y_1}(x)| \leqq k|y_1-y_2| .$$

Consequently,

$$(10) \quad |y_1-y_2| \leqq \tfrac{1}{2} R \Longrightarrow \left| \int_B g(1-\psi_{y_2}) d(\lambda^G_{y_1} - \lambda^G_{y_2}) \right| \leqq$$

$$\leqq k|y_1-y_2| \, \mathcal{H}_{m-1}(\hat{B}) ,$$

where $\hat{B} = \{y \in B;\ |n^G(y)| > 0\}$ is the reduced boundary of G (cf. 2.14). Combining (8) - (10) we arrive at

$$|y_1-y_2| \leqq \tfrac{1}{2} R \Longrightarrow |Vg(y_1) - Vg(y_2)|$$

$$\leqq |y_1-y_2| \cdot (v^G + k\,\mathcal{H}_{m-1}(\hat{B}))$$

which shows that all the functions in (2) are equicontinuous. Since, in view of (7), all these functions are also uniformly bounded, we conclude that $V \in \mathcal{C}$.

Let $\alpha \in R^1$ and denote by δ_y the Dirac measure at y. Then

$$\|W^G_B - \alpha I - V\| =$$

$$= \sup_{y \in B} \sup_{\substack{g \in \mathcal{C}(B) \\ \|g\| \leq 1}} (W_B^G - \alpha I - V)g(y) =$$

$$= \sup_{y \in B} \int_B \psi_y d\,|\lambda_y^G - \alpha \delta_y| \,.$$

Note that, in view of (1) and the isoperimetric lemma stated in 2.14, $\hat{B}$ is dense in B. Since the function

$$y \longmapsto \int_B \psi_y \, d\,|\lambda_y^G - \alpha \delta_y|$$

is (as a supremum of a family of continuous functions) lower--semicontinuous on B we get with help of (13) in 2.8 and (15) in 1.11 the estimates

$$\|W_B^G - \alpha I - V^R\| = \sup_{y \in \hat{B}} \int_B \psi_y d\,|\lambda_y^G - \alpha \delta_y| \leq$$

$$\leq \sup_{y \in \hat{B}} \left[|\lambda_y^G - \alpha \delta_y|(\{y\}) + |\lambda_y^G|(\Omega_{2R}(y) \setminus \{y\}) \right] =$$

$$= |\tfrac{1}{2} - \alpha| + \sup_{y \in \hat{B}} v_{2R}^G(y) \leq |\tfrac{1}{2} - \alpha| + \sup_{y \in B} v_{2R}^G(y) \,.$$

Noting that $V^R \in \mathcal{V}$ and making $R \to 0+$ we get

$$(11) \qquad \omega(W_B^G - \alpha I) \leq |\tfrac{1}{2} - \alpha| + v_o^G \,.$$

Now we are going to establish the lower estimate of $\omega(W_B^G - \alpha I)$. By the Radon theorem, any $V \in \mathcal{V}$ can be arbitrarily closely approximated by finite dimensional operators of the form

$$\hat{V}f = \sum_{k=1}^{q} \langle f, \nu_k \rangle \varphi_k$$

with $\varphi_k \in \mathcal{C}(B)$ and $\nu_k \in \mathcal{C}'(B)$. Any ν_k can be arbitrarily closely approximated (in the norm of $\mathcal{C}'(B)$) by a $\bar{\nu}_k \in \mathcal{C}'(B)$ with the property that the set

(12) $$\{y \in B;\ |\bar{\nu}_k|(\{y\}) > 0\}$$

is finite. Defining

(13) $$\bar{V} \ldots = \sum_{k=1}^{q} \langle \ldots, \bar{\nu}_k \rangle \varphi_k$$

we see that $\|V-\bar{V}\|$ can be made as small as we want. It follows from these observations that

(14) $$\omega(W_B^G - \alpha I) = \inf_{\bar{V}} \|W_B^G - \alpha I - \bar{V}\|,$$

where $\bar{V}$ ranges over all finite dimensional operators of the form (13) such that each of the sets (12) is finite. Let us fix such a $\bar{V}$ and denote by D the (finite) set of all $y \in B$ that are charged by $\sum_k |\bar{\nu}_k|$ (so that D is the union of all the sets (12)). Every $\bar{\nu}_k$ splits into $\nu_k^1 \in \mathcal{C}'(B)$ not charging singletons and a finite combination of Dirac measures, to be denoted by ν_k^2. Since y is the only point that can be charged by $\lambda_y^G - \alpha\delta_y$, we have for $y \in B \setminus D$

$$\|\lambda_y^G - \alpha\delta_y - \sum_k \varphi_k(y)\bar{\nu}_k\| = \|\lambda_y^G - \alpha\delta_y - \sum_k \varphi_k(y)\nu_k^1\| +$$

$$+ \| \sum_k \varphi_k(y) \nu_k^2 \| ,$$

whence

$$\| W_B^G - \alpha I - \bar{V} \| \geq \sup_{y \in B \setminus D} \| \lambda_y^G - \alpha \delta_y - \sum_k \varphi_k(y) \nu_k^1 \| =$$

$$= \sup_{y \in B} \| \lambda_y^G - \alpha \delta_y - \sum_k \varphi_k(y) \nu_k^1 \| ,$$

because $B \setminus D$ is dense in B (there are no isolated points in B by (1)) and because

$$y \longmapsto \| \lambda_y^G - \alpha \delta_y - \sum_k \varphi_k(y) \nu_k^1 \| =$$

$$= \sup_{\|g\| \leq 1} \left[(W_B^G - \alpha I) g(y) - \sum_k \langle g, \nu_k^1 \rangle \varphi_k(y) \right]$$

is lower-semicontinuous on B. If $\max_k \| \varphi_k \| = M$, we get for any $r > 0$

$$\sup_{y \in B} \| \lambda_y^G - \alpha \delta_y - \sum_k \varphi_k(y) \nu_k^1 \| \geq$$

$$\geq \sup_{y \in B} | \lambda_y^G - \alpha \delta_y - \sum_k \varphi_k(y) \nu_k^1 | (\Omega_r(y)) \geq$$

$$\geq \sup_{y \in B} | \lambda_y^G - \alpha \delta_y | (\Omega_r(y)) - M \sup_{y \in B} \sum_k | \nu_k^1 | (\Omega_r(y)) .$$

According to 2.8, 2.9, 1.11 we have

$$| \lambda_y^G - \alpha \delta_y | (\Omega_r(y)) = | d_G(y) - \alpha | + v_r^G(y) .$$

Since $\sum_k | \nu_k^1 |$ does not charge singletons,

$$\lim_{r \to 0+} \sup_{y \in B} \sum_k |\nu_k^1|(\Omega_r(y)) = 0 .$$

Making $r \to 0+$ in the above estimates we obtain

$$\| W_B^G - \alpha I - \bar{V} \| \geqq \lim_{r \to 0+} \sup_{y \in B} \{ |d_G(y) - \alpha| + v_r^G(y) \} .$$

Since $\bar{V}$ was arbitrarily chosen we conclude from (14) that

$$\omega(W_B^G - \alpha I) \geqq \lim_{r \to 0+} \sup_{y \in B} \{ |d_G(y) - \alpha| + v_r^G(y) \} .$$

Combining this with (11) we arrive at

$$\lim_{r \to 0+} \sup_{y \in B} \{ |d_G(y) - \alpha| + v_r^G(y) \} \leqq \omega(W_B^G - \alpha I) \leqq$$

$$\leqq |\tfrac{1}{2} - \alpha| + V_o^G .$$

In order to prove (6) it remains to observe that the left wing in this string of inequalities is greater or equal to the right wing. Indeed, for any $r > 0$

$$\sup_{y \in B} \{ |d_G(y) - \alpha| + v_r^G(y) \} \geqq \sup_{y \in \hat{B}} \{ \dots \} =$$

$$= |\tfrac{1}{2} - \alpha| + \sup_{y \in \hat{B}} v_r^G(y) .$$

Since $\hat{B}$ is dense in B and the function

$$y \longmapsto v_r^G(y)$$

is lower-semicontinuous on B (cf. 1.11), we have

$$\sup_{y \in \hat{B}} v_r^G(y) = \sup_{y \in B} v_r^G(y) ,$$

so that

$$\lim_{r\to 0+} \sup_{y\in B} \{ |d_G(y) - \alpha| + v_r^G(y)\} \geq |\tfrac{1}{2} - \alpha| + V_o^G$$

and the proof is complete.

Remark. The equations

$W_B^G f = g$ ($g \in \mathcal{C}(B)$ prescribed, $f \in \mathcal{C}(B)$ unknown),

$(W_B^G)' \mu = \nu$ ($\nu \in \mathcal{C}'(B)$ prescribed, $\mu \in \mathcal{C}'(B)$ unknown)

(which, as we observed in §§ 2, 3, are connected with the Dirichlet and the Neumann problem) may be written in the form of the equations of the second kind

$$\left[I + \frac{1}{\alpha} (W_B^G - \alpha I)\right] f = \frac{1}{\alpha} g ,$$

$$\left[I + \frac{1}{\alpha} (W_B^G - \alpha I)\right]' \mu = \frac{1}{\alpha} \nu ,$$

for which the Fredholm theory has been developed. In connection with applicability of the Riesz-Schauder version of this theory to the above equations it is important to know whether the parameter $\alpha \neq 0$ can be chosen in such a way that

(15) $$\omega\left(\frac{1}{\alpha} W_B^G - I\right) < 1 .$$

This question is answered by the following corollary of 4.1.

4.2. Corollary. The inequality (15) holds for suitable $\alpha \neq 0$ if and only if

(16) $$V_o^G < \frac{1}{2} .$$

If (16) holds, then the minimal value of

$$\omega(\frac{1}{\alpha} W_B^G - I)$$

is attained for $\alpha = \frac{1}{2}$, when $\frac{1}{\alpha} W_B^G - I = 2W_B^G - I$ reduces to the Neumann operator T^G (cf. (51)):

$$\min_{\alpha \neq 0} \omega(\frac{1}{\alpha} W_B^G - I) = \omega T^G = 2V_o^G .$$

Proof. We have by (6)

$$\omega(\frac{1}{\alpha} W_B^G - I) = \frac{1}{|\alpha|} (|\frac{1}{2} - \alpha| + V_o^G) \tag{17}$$

and elementary discussion shows that this quantity can be less than 1 only if $V_o^G < \frac{1}{2}$, when the minimal value of (15) is attained for $\alpha = \frac{1}{2}$ and equals $2V_o^G$.

4.3. Observation. If (16) holds, then necessarily

$$\mathcal{H}_m(B) = 0 . \tag{18}$$

Proof. Using (6) with $\alpha = \frac{1}{2}$ we get

$$\sup_{y \in B} |d_G(y) - \tfrac{1}{2}| \leqq \lim_{r \to 0+} \sup_{y \in B} (|d_G(y) - \tfrac{1}{2}| + v_r^G(y)) =$$
$$= V_o^G < \frac{1}{2} ,$$

so that, for suitable $\varepsilon > 0$,

$$y \in B \Longrightarrow \varepsilon < d_G(y) < 1 - \varepsilon . \tag{19}$$

It remains to recall that the density of a Lebesgue measurable $G \subset R^m$ cannot be different both from 0 and 1 on a set of positive $\mathcal{H}_m$-measure.

4.4. Remark. In what follows we always assume (18).

We have so far considered only real-valued functions and measures. For our next investigation it will be useful to admit complex-valued functions and measures. From now on we shall mean by $\mathcal{C}(B)$ the Banach space of all continuous complex-valued functions f on B normed by $\sup_{y \in B} |f(y)| = \|f\|$. $\mathcal{C}'(B)$ will always mean the Banach space of all countably additive complex-valued Borel measures μ on R^m with support in B, the norm being given by the total variation of μ on R^m, $\|\mu\| = |\mu|(R^m)$; $\mathcal{C}'(B)$ will be identified with the dual space to $\mathcal{C}(B)$. For complex $\mu \in \mathcal{C}'(B)$ the potential $\mathcal{U}\mu$ is defined as before and, for the case when G is open, its generalized normal derivative $N^G\mathcal{U}\mu$ may be considered as in §1. The double layer potentials W^Gf investigated in §2 can naturally be defined for complex-valued f. The reader will easily observe that the basic results established before remain in force for the complex case. Under the assumption $V^G < +\infty$, which we adopt throughout the rest of this text, the operator W^G_B (sending $f \in \mathcal{C}(B)$ into the restriction to B of the corresponding double layer potential W^Gf) rests bounded on $\mathcal{C}(B)$ and for open G we have again

$$(W^G_B)' = N^G\mathcal{U} .$$

In accordance with §2, $\mathcal{C}'_c(B)$ will stand for the subspace of those complex $\mu \in \mathcal{E}'(B)$ for which there exists a continuous function $\mathcal{U}_c\mu$ on R^m coinciding with $\mathcal{U}\mu$ on $R^m \setminus B$; the restriction of $\mathcal{U}_c\mu$ to B will be denoted by $\mathcal{U}_o\mu$.

We shall say that W is a Plemelj's operator if W is a bounded linear operator acting on $\mathcal{C}(B)$ whose dual W' maps $\mathcal{E}_c'(B)$ into itself and

$$\mu \in \mathcal{E}_c'(B) \Rightarrow W(\mathcal{U}_o\mu) = \mathcal{U}_o(W'\mu) .$$

We shall now consider the family of operators

(20) $$T_\gamma^G \equiv T_\gamma = T^G + \gamma I ,$$

where $T^G = 2W_B^G - I$ is the Neumann operator and the parameter γ runs over the set $\mathbb{C}$ of all complex numbers. It follows from Plemelj's exchange theorem established in §2 (extended to the complex case) that all the operators in (20) are Plemelj's. Put

(21) $$\Omega = \{\gamma \in \mathbb{C} ; |\gamma| > \omega T^G\} .$$

It is well known (cf. [RS]) that there is a countable set $\mathcal{N} \subset \Omega$ consisting of isolated points such that, for each $\gamma \in \Omega \setminus \mathcal{N}$, the inverse operators T_γ^{-1} and $(T_\gamma')^{-1}$ are defined on $\mathcal{C}(B)$ and $\mathcal{C}'(B)$, respectively. We are now going to prove that also the operators T_γ^{-1} $(\gamma \in \Omega \setminus \mathcal{N})$ are Plemelj's. This will permit us later to prove that, for each $\gamma \in \Omega$, any $\mu \in \mathcal{C}'(B)$ satisfying the homogeneous equation $T_\gamma'\mu = 0$ necessarily belongs to $\mathcal{E}_c'(B)$, which is a fact important in connection with treating the boundary value problems.

4.5. Lemma. Let $\mu_n \in \mathcal{E}_c'(B)$ $(n = 1,2,\ldots)$, $\sum_n \|\mu_n\| < +\infty$, $\sum_n \|\mathcal{U}_o \mu_n\| < +\infty$. Then $\mu = \sum_n \mu_n \in$ $\in \mathcal{E}_c'(B)$ and

$$\mathcal{U}_o \mu = \sum_n \mathcal{U}_o \mu_n .$$

Proof. Fix $R > 0$ such that $B \subset \Omega_R(0)$. Then the series $\sum_n \mathcal{U}\mu_n$ is uniformly convergent on $\partial \Omega_R(0)$ so that the series of (complex-valued) harmonic functions

$$(22) \qquad \sum_n \mathcal{U}\mu_n ,$$

considered on $\Omega_R(0) \setminus B = D$, consists of functions continuously extendable from D to $\operatorname{cl} D$ whose continuous extensions form a uniformly convergent series on the boundary ∂D. Consequently, (22) is uniformly convergent on D and also the sum

$$\sum_n \mathcal{U}\mu_n = \mathcal{U}\mu$$

is continuously extendable to B. The rest is obvious.

4.6. Lemma. All the operators T_α with $|\alpha| > \|T^G\|$ have Plemelj's inverses. If T_β^{-1} is a Plemelj's operator with $\|T_\beta^{-1}\| \leq K$, then also all the operators T_γ with $|\gamma - \beta| < 1/K$ possess Plemelj's inverses.

Proof. If T_β has a bounded inverse with $\|T_\beta^{-1}\| \leq K$, then we conclude from the identity

$$T_\gamma = T_\beta + (\gamma - \beta)I = T_\beta\left[I - (\beta - \gamma)T_\beta^{-1}\right]$$

that for any γ with $|\gamma - \beta| < 1/K$ the inverse operators to T_γ and T'_γ are given by

$$T_\gamma^{-1} = \sum_{n=0}^{\infty} (\beta - \gamma)^n (T_\beta^{-1})^{n+1}$$

and

$$(T'_\gamma)^{-1} = \sum_{n=0}^{\infty} (\beta - \gamma)^n \left[(T'_\beta)^{-1}\right]^{n+1} ,$$

respectively. If T_β^{-1} is Plemelj's and $\mu \in \mathcal{E}'_c(B)$, we have for each n

$$\mathcal{U}_0 \left[(T'_\beta)^{-1}\right]^{n+1} \mu = (T_\beta^{-1})^{n+1} \mathcal{U}_0 \mu ,$$

so that

$$\|\mathcal{U}_0 \{(\beta - \gamma)^n \left[(T'_\beta)^{-1}\right]^{n+1} \mu\}\| \leqq |\beta - \gamma|^n \|T_\beta^{-1}\|^{n+1} \|\mathcal{U}_0 \mu\| \leqq$$

$$\leqq |\beta - \gamma|^n K^{n+1} \|\mathcal{U}_0 \mu\|$$

and

$$\sum_{n=0}^{\infty} \|\mathcal{U}_0 \{(\beta - \gamma)^n \left[(T'_\beta)^{-1}\right]^{n+1} \mu\}\| < +\infty$$

provided $|\beta - \gamma| < K$. Employing 4.5 and noting that $\sum_{n=0}^{\infty} |\beta - \gamma|^n \|(T'_\beta)^{-1}\|^{n+1} < +\infty$ we conclude for these γ that $(T'_\gamma)^{-1}\mu \in \mathcal{E}'_c(B)$ and

$$\mathcal{U}_0 \left[(T'_\gamma)^{-1}\mu\right] = \sum_{n=0}^{\infty} \mathcal{U}_0 \{(\beta - \gamma)^n \left[(T'_\beta)^{-1}\right]^{n+1}\}\mu =$$

$$= \sum_{n=0}^{\infty} (\beta - \gamma)^n (T_\beta^{-1})^{n+1} \mathcal{U}_0 \mu = T_\gamma^{-1}(\mathcal{U}_0 \mu) ,$$

so that T_γ^{-1} is Plemelj's.

Let us write simply $T \equiv T^G$ and consider $|\alpha| > \|T\|$. Then

$$T_\alpha^{-1} = (T + \alpha I)^{-1} = -\sum_{n=0}^{\infty} (-\alpha)^{-(n+1)} T^n ,$$

$$(T_\alpha')^{-1} = -\sum_{n=0}^{\infty} (-\alpha)^{-(n+1)} (T')^n .$$

Using the fact that T is a Plemelj's operator we get for $\mu \in \mathcal{E}_c'(B)$ and each n that

$$(T')^n \mu \in \mathcal{E}_c'(B) , \quad \mathcal{U}_0 [(T')^n \mu] = T^n(\mathcal{U}_0 \mu) .$$

Employing 4.5 and reasoning as above we obtain that $(T_\alpha')^{-1}\mu \in \mathcal{E}_c'(B)$ and

$$\mathcal{U}_0 [(T_\alpha^{-1})'\mu] = -\sum_{n=0}^{\infty} (-\alpha)^{-(n+1)} \mathcal{U}_0 [(T')^n \mu] =$$

$$= -\sum_{n=0}^{\infty} (-\alpha)^{-(n+1)} T^n(\mathcal{U}_0 \mu) = T_\alpha^{-1}(\mathcal{U}_0 \mu) ,$$

so that T_α^{-1} is Plemelj's.

<u>4.7. Lemma.</u> Define Ω by (21) and denote by $\mathcal{N}$ the set of all $\alpha \in \Omega$ for which the equation $T_\alpha f = 0$ has non-trivial solutions $f \in \mathcal{E}(B)$. Then all the operators T_γ with $\gamma \in \Omega \setminus \mathcal{N}$ possess inverses that are Plemelj's.

Proof. Let us recall that $\mathcal{N}$ (which is just the set of those $\alpha \in \Omega$ for which T_α does not possess a bounded inverse on $\mathcal{C}(B)$) consists of isolated points (cf. [RS]) so that $\Omega \setminus \mathcal{N}$ is an open connected set. Let Ω_0 be the set of all $\gamma \in \Omega \setminus \mathcal{N}$ for which T_γ^{-1} is a Plemelj's operator. We know from lemma 4.6 that Ω_0 is open and non-void so that it will suffice to show that Ω_0 is relatively closed in $\Omega \setminus \mathcal{N}$ in order to prove $\Omega_0 = \Omega \setminus \mathcal{N}$. Consider an arbitrary $\gamma \in (\Omega \setminus \mathcal{N}) \cap \mathrm{cl}\, \Omega_0$. Since the map

$$\beta \longmapsto T_\beta^{-1}$$

is continuous (from $\Omega \setminus \mathcal{N}$ to the space of all bounded linear operators acting on $\mathcal{C}(B)$), there is a closed ball $H = \mathrm{cl}\, \Omega_r(\gamma) \subset \Omega \setminus \mathcal{N}$ centered at γ and a $K > 0$ such that

$$\beta \in H \Longrightarrow \| T_\beta^{-1} \| \leq K .$$

Now choose a $\beta \in H \cap \Omega_0$ with $|\beta - \gamma| < \frac{1}{K}$. Lemma 4.6 implies that $\gamma \in \Omega_0$ so that Ω_0 is (relatively) closed in $\Omega \setminus \mathcal{N}$ and the proof is complete.

Remark. Our next aim is to prove that, for any positive integer p and any $\gamma \in \Omega$, every $\mu \in \mathcal{C}'(B)$ satisfying the homogeneous equation

$$(23) \qquad (T_\gamma')^p \mu = 0$$

necessarily belongs to $\mathcal{C}_c'(B)$. We shall start with several simple observations and adopt the following notation. If Q is a linear operator acting on a Banach space, we denote by $\mathcal{N}(Q)$ its null-space and by $\mathcal{R}(Q)$ its range; dim S will be used to denote the (algebraic) dimension of a linear space S .

4.8. Lemma. If Q is a linear operator with $\dim \mathcal{N}(Q) < +\infty$, then $\dim \mathcal{N}(Q^p) < +\infty$ for every positive integer p .

Proof. Let $p > 1$ and suppose that the assertion has been verified for $(p-1)$. Put $\tilde{Q} = Q^{p-1}$ and choose a basis $\{y_1,\dots,y_r\}$ in $\mathcal{N}(\tilde{Q})$ and a basis $\{z_1,\dots,z_s\}$ in $\mathcal{N}(Q) \cap \cap \mathcal{R}(\tilde{Q})$. Further choose x_i with $\tilde{Q}x_i = z_i$ $(i = 1,\dots,s)$ and denote by X the linear space spanned by $\{x_1,\dots,x_s, y_1,\dots,y_r\}$. We are going to prove that $\mathcal{N}(Q^p) \subset X$. Let $x_o \in \mathcal{N}(Q^p)$. Since $Q(\tilde{Q}x_o) = 0$, we have $\tilde{Q}x_o = \sum_{i=1}^{s} \alpha_i z_i$

for suitable scalars α_i , so that $\tilde{x} = x_o - \sum_{i=1}^{s} \alpha_i x_i$

satisfies $\tilde{Q}\tilde{x} = 0$, whence $\tilde{x} = \sum_{j=1}^{r} \beta_j y_j$ for suitable scalars β_j . We see that $x_o = \sum_{i=1}^{s} \alpha_i x_i + \sum_{j=1}^{r} \beta_j y_j \in X$ and $\dim \mathcal{N}(Q^p) \leqq r+s$.

4.9. Lemma. Let p be a positive integer, $\lambda \in \Omega$ and suppose that $f_1,\dots,f_q \in \mathcal{C}(B)$ are linearly independent solutions of the equation

(24) $$T_\lambda^p f = 0 .$$

Then there exist $\mu_1, \ldots, \mu_q \in \mathcal{E}_c'(B)$ such that

$$\langle f_i, \mu_j \rangle = \delta_{ij} \quad (= \text{Kronecker's symbol}), \quad 1 \leqq i, \ j \leqq q .$$

Proof. We shall first show that, given an arbitrary non-zero element $f \in \mathcal{E}(B)$, there is always a $\mu \in \mathcal{E}_c'(B)$ with $\langle f, \mu \rangle \neq 0$ (so that the assertion is obviously valid for $q = 1$ and the rest will then follow by an inductive argument). Fix $y \in B$ with $f(y) \neq 0$ and choose $r > 0$ small enough to guarantee that $\operatorname{Re} \frac{f(x)}{f(y)} \geqq \frac{1}{2}$ for all $x \in E = B \cap \Omega_r(y)$. Writing $n^G(.)$ for the vector-valued function introduced in 2.14 and g for the characteristic function of E on B we define $\mu \in \mathcal{E}'(B)$ by

$$d\mu(x) = g(x)\,|n^G(x)|\, d\mathcal{H}_{m-1}(x) , \quad x \in B .$$

It follows from the isoperimetric lemma stated in 2.14 and our assumption (1) that $\mu(B) = \mu(E) > 0$ and our construction guarantees that $|\langle f, \mu \rangle| \geqq \frac{1}{2} |f(y)|\, \mu(E) > 0$. Lemma 2.18 assures that the potential $\mathcal{U}\mu$ is continuous on the whole space R^m, so that $\mu \in \mathcal{E}_c'(B)$.

Let now $q > 1$ and suppose that the lemma has been verified for $q-1$, so that we may fix $\{\mu_2', \ldots, \mu_q'\}$ in $\mathcal{E}_c'(B)$ forming a biorthogonal system with $\{f_2, \ldots, f_q\}$. If $\tilde{\mu} \in \mathcal{E}_0'(B)$, then the measure

$$(25) \qquad \tilde{\mu} - \sum_{k=2}^{q} \langle f_k, \tilde{\mu} \rangle \mu_k'$$

belongs to $\mathcal{E}_c'(B)$ and is orthogonal to $f_2, \ldots, f_q$. The orthogonality of (25) to f_1 would mean that

$$\langle f_1, \tilde{\mu} \rangle = \sum_{k=2}^{q} \langle f_k, \tilde{\mu} \rangle \langle f_1, \mu_k' \rangle$$

so that, with $c_k = \langle f_1, \mu_k' \rangle$,

$$\langle f_1 - \sum_{k=2}^{q} c_k f_k, \tilde{\mu} \rangle = 0 . \tag{26}$$

In view of linear independence of $\{f_1,\dots,f_q\}$ we may fix $\tilde{\mu} \in \mathscr{E}_c'(B)$ for which (26) does not hold and normalize it in such a way that the corresponding $\mu_1 = \tilde{\mu} - \sum_{k=2}^{q} \langle f_k, \tilde{\mu} \rangle \mu_k'$ satisfies $\langle f_1, \mu_1 \rangle = 1$; as all elements of the form (25), μ_1 is orthogonal to $f_2,\dots,f_q$. The same reasoning provides the remaining μ_j for $j = 2,\dots,q$.

<u>4.10. Theorem.</u> If p is a positive integer and $\gamma \in \Omega$, then any $\mu \in \mathscr{E}'(B)$ satisfying the homogeneous equation (23) necessarily belongs to $\mathscr{E}_c'(B)$.

<u>Proof.</u> The assertion is obvious in case $\gamma \in \Omega \setminus \mathscr{N}$, because T_γ' is invertible for such γ so that any $\mu \in \mathscr{E}'(B)$ satisfying (23) must be trivial. We may therefore assume that $\gamma \in \mathscr{N}$. It is well known that the resolvents of the operators T_α , T_α' have poles at γ (cf. [RS]) and these poles are of the same order (cf. chap. VIII in [Y]), say p_0 . We may clearly suppose that $p \geqq p_0$. Fix $r > 0$ small enough to guarantee that the closed disc $H = \mathrm{cl}\, \Omega_r(\gamma)$

is contained in Ω and

$$H \cap \mathcal{N} = \{\gamma\}.$$

Let C denote the counterclockwise oriented circumference ∂H and define the operator A_{-1} acting on $\mathcal{E}(B)$ by

$$A_{-1} = \frac{1}{2\pi i} \int_C T_\alpha^{-1} \, d\alpha \tag{27}$$

(cf. chap. VIII in [Y]). We are going to prove that A_{-1} is a Plemelj's operator. Note that the function

$$\alpha \longmapsto T_\alpha^{-1}$$

is continuous on ∂H, so that the integral (27) represents a limit of the Riemann's sums S_n, each S_n being a finite linear combination (with complex coefficients) of the operators $T_{\alpha_j}^{-1}$ with $\alpha_j \in \partial H$. In view of 4.7, each S_n is a Plemelj's operator. Passing to subsequences, if necessary, we may clearly achieve that

$$\sum_{n=1}^{\infty} \|S_n - S_{n+1}\| < +\infty .$$

Put $R_1 = S_1$, $R_{n+1} = S_{n+1} - S_n$ $(n = 1,2,\dots)$, so that

$$A_{-1} = \sum_{n=1}^{\infty} R_n ,$$

each R_n being a Plemelj's operator. Consider now a $\mu \in \mathcal{E}_c'(B)$. Then $\mu_n = R_n' \mu \in \mathcal{E}_c'(B)$ and $\mathcal{U}_o \mu_n = R_n \mathcal{U}_o \mu$, so that $\|\mathcal{U}_o \mu_n\| \leq \|R_n\| \, \|\mathcal{U}_o \mu\|$ and, consequently,

$$\sum_{n=1}^{\infty} \| \mathcal{U}_0 \mu_n \| < +\infty .$$

In view of

$$\sum_{n=1}^{\infty} \| \mu_n \| \leqq \left(\sum_{n=1}^{\infty} \| R_n \| \right) \| \mu \| < +\infty$$

we conclude from lemma 4.5 that $A'_{-1}\mu = \sum_{n=1}^{\infty} \mu_n \in \mathcal{E}'_c(B)$ and

$$\mathcal{U}_0(A'_{-1}\mu) = \sum_{n=1}^{\infty} \mathcal{U}_0(R'_n\mu) = \sum_{n=1}^{\infty} R_n(\mathcal{U}_0\mu) = A_{-1}(\mathcal{U}_0\mu) ,$$

so that A_{-1} is a Plemelj's operator. For its dual we have a representation analoguous to (27):

(28) $$A'_{-1} = \frac{1}{2\pi i} \int_C (T_\alpha^{-1})' \, d\alpha$$

(cf. [Y], chap. VIII, 7). Further we have

(29) $$\mathcal{R}(A'_{-1}) = \{ \mu \in \mathcal{E}'(B); \ (T'_\gamma)^p \mu = 0 \} ,$$

(30) $$\mathcal{R}(A_{-1}) = \{ f \in \mathcal{E}(B); \ T_\gamma^p f = 0 \}$$

(cf. [Y], chap. VIII, 8), $\dim \mathcal{R}(A_{-1}) < +\infty$ (cf. 4.8). Let $\{f_1, \dots, f_q\}$ be a basis in (30). Then A_{-1} has the form

$$A_{-1} \dots = \sum_{k=1}^{q} \langle \dots, \mu'_k \rangle f_k ,$$

where $\mu'_k \in \mathcal{E}'(B)$ $(k = 1, \dots, q)$. Consequently,

$$(31) \qquad A'_{-1} \cdots = \sum_{k=1}^{q} \langle f_k, \cdots \rangle \mu_k \, .$$

Employing lemma 4.9 we choose $\mu_k \in \mathcal{E}'_c(B)$ such that $\langle f_j, \mu_k \rangle = \delta_{jk}$, $1 \leqq j$, $k \leqq q$. We conclude from (31) that $A'_{-1}\mu_k = \mu'_k$, so that $\mu'_k \in \mathcal{E}'_c(B)$ $(k = 1,\ldots,q)$, because A_{-1} is a Plemelj's operator. We have thus $\mathcal{R}(A'_{-1}) \subset \subset \mathcal{E}'_c(B)$ which together with (29) completes the proof.

§ 5

Boundary value problems

As in the preceding paragraph we assume that $G \subset R^m$ is a Borel set with a compact boundary B and $C = R^m \setminus G$; we again always suppose that each ball centered at a point in B meets both G and C in a set of positive $\mathcal{H}_m$-measure.

We observed in 2.24 that the Dirichlet problem corresponding to $\operatorname{int} C$ and the boundary condition $g \in \mathcal{C}(B)$ leads naturally (at least when $B \subset \operatorname{cl}(\operatorname{int} C)$) to the equation $W_B^G f = g$ for an unknown $f \in \mathcal{C}(B)$ (cf. (49) in §2) and that the generalized Neumann problem for G and the prescribed boundary condition $\mu \in \mathcal{C}'(B)$ consists (when G is open) in determining a $\nu \in \mathcal{C}'(B)$ satisfying the dual equation $(W_B^G)' \nu = \mu$ (cf. (50) in §2); we have also noticed that these equations may conveniently be transformed into the form

$$\left[I + \frac{1}{\alpha}(W_B^G - \alpha I)\right] f = \frac{1}{\alpha} f ,$$

$$\left[I + \frac{1}{\alpha}(W_B^G - \alpha I)\right]' \nu = \frac{1}{\alpha} \mu ,$$

where $\alpha \neq 0$ is a parameter. We know from §4 that the minimal distance of the family of operators

$$\frac{1}{\alpha}(W_B^G - \alpha I) \equiv Q_\alpha$$

from the subspace of all compact operators on $\mathcal{C}(B)$ is less than 1 if and only if

(1) $$V_o^G < \frac{1}{2}$$

and that the optimal choice of the parameter is $\alpha = \frac{1}{2}$ (cf. also 2.25); for this choice the minimum of ωQ_α is attained, Q_α reduces to $2(W_B^G - \frac{1}{2} I) = T^G \equiv T$ which is the Neumann operator and the above equations take the form

(2) $$(I + T)f = g_1 ,$$

(3) $$(I + T)'\nu = \mu_1$$

with $g_1 = 2g$, $\mu_1 = 2\mu$.

Our first objective in this paragraph will be the investigation of the corresponding homogeneous equations

(2_o) $$(I + T)f = 0 \quad (\in \mathcal{C}(B)) ,$$

(3_o) $$(I + T)'\nu = 0 \quad (\in \mathcal{C}'(B))$$

under the assumption (1).

5.1. Remark. If K is a bounded linear operator acting on $\mathcal{C}(B)$ and $y \in B$, then the Riesz representation theorem guarantees that there is a uniquely determined $\varkappa_y \in \mathcal{C}'(B)$ such that

$$Kf(y) = \int_B f d\varkappa_y , \quad f \in \mathcal{C}(B) .$$

Let us now denote by $\mathcal{B}(B) \equiv \mathcal{B}$ the Banach space of all bounded Baire functions h on B equipped with the norm $\|h\| = \sup|h|(B)$. For $h \in \mathcal{B}$ and $y \in B$ we may again define $Kh(y) = \int_B h d\varkappa_y$; it is easily observed that $y \mapsto Kh(y)$

is a bounded Baire function. Indeed, let $\mathcal{B}_1$ denote the class of all $h_1 \in \mathcal{B}$ with $\|h_1\| \leqq 1$, $Kh_1 \in \mathcal{B}$ and $\|Kh_1\| \leqq \|K\|$. It is easy to see that $\mathcal{B}_1$ contains the limit of each pointwise convergent sequence of its elements. Since $\{f \in \mathcal{C}(B);\ \|f\| \leqq 1\} \subset \mathcal{B}_1$ we conclude that $\mathcal{B}_1$ coincides with the whole closed unit ball in $\mathcal{B}$ and $\|K\|_{\mathcal{B}}$, the norm of the operator $K : h \to Kh$ acting on $\mathcal{B}$, is just the same as $\|K\|$. This observation shows that each bounded linear operator K acting on $\mathcal{C}(B) \equiv \mathcal{C}$ extends naturally to a bounded linear operator on $\mathcal{B}$; we shall denote the extended operator by the same symbol. The symbol K', however, will always denote the dual operator acting on $\mathcal{C}'(B) \equiv \mathcal{C}'$. In order to avoid confusion we shall write

$$\mathcal{N}_{\mathcal{C}}(K) = \{f \in \mathcal{C};\ Kf = 0 \in \mathcal{C}\},$$

$$\mathcal{N}_{\mathcal{B}}(K) = \{h \in \mathcal{B};\ Kh = 0 \in \mathcal{B}\},$$

$$\mathcal{N}_{\mathcal{C}'}(K') = \{\nu \in \mathcal{C}';\ K'\nu = 0 \in \mathcal{C}'\},$$

$$\mathcal{R}_{\mathcal{B}}(K) = \{Kh;\ h \in \mathcal{B}\}, \quad \mathcal{R}_{\mathcal{C}}(K) = \{Kf;\ f \in \mathcal{C}\},$$

$$\mathcal{R}_{\mathcal{C}'}(K') = \{K'\nu\ ;\ \nu \in \mathcal{C}'\}$$

etc. These observations apply, in particular, to the operators $W_B^G \equiv W$, $T^G \equiv T$ which we have so far considered on $\mathcal{C}$ only.

We shall now recall a version of Fredholm's alternative which will be useful in connection with treating boundary value problems with help of the equations (2), (3).

5.2. Lemma. Let $K = I + Q$, where I is the identity operator and Q is a bounded linear operator on $\mathcal{C}$ with $\omega Q < 1$. Then

(4) $$\dim \mathcal{N}_{\mathcal{B}}(K) = \dim \mathcal{N}_{\mathcal{C}'}(K') < +\infty$$

and

(5) $$\mathcal{R}_{\mathcal{B}}(K) = \{g \in \mathcal{B} \,;\, \langle g, \nu \rangle = 0 \text{ for all } \nu \in \mathcal{N}_{\mathcal{C}'}(K')\},$$
$$\mathcal{R}_{\mathcal{C}}(K) = \{f \in \mathcal{C} \,;\, \langle f, \nu \rangle = 0 \text{ for all } \nu \in \mathcal{N}_{\mathcal{C}'}(K')\},$$
$$\mathcal{R}_{\mathcal{C}'}(K') = \{\mu \in \mathcal{C}' ;\, \langle f, \mu \rangle = 0 \text{ for all } f \in \mathcal{N}_{\mathcal{C}}(K)\}.$$

__Proof.__ Let us recall that any compact operator acting on $\mathcal{C}$ can be arbitrarily closely approximated by finite dimensional operators. We have thus

$$Q = Q_o + Q_1 ,$$

where $\|Q_1\| < 1$ and Q_o is a finite dimensional operator of the form

$$Q_o \ldots = \sum_{i=1}^{q} \langle \ldots, \mu_i \rangle f_i$$

with $\mu_i \in \mathcal{C}'$ and $f_i \in \mathcal{C}$ $(i = 1,\ldots,q)$. Consider now the equation

(6) $$Kf = g .$$

Multiplying it by $(I + Q_1)^{-1} = \sum_{k=0}^{\infty} (-1)^k Q_1^k$ (where $Q_1^o = I$) we transform it into

(7) $$(I + \tilde{Q})f = h ,$$

where $\tilde{Q} = (I + Q_1)^{-1} Q_o$, $h = (I + Q_1)^{-1} g$. Notice that $\mathcal{C}'$ (although it represents only a proper subspace of the dual space to $\mathcal{B}$) is total over $\mathcal{B}$ in the sense that

$$(h \in \mathcal{B} , \langle h, \mu \rangle = 0 \text{ for all } \mu \in \mathcal{C}') \Rightarrow h = 0 \in \mathcal{B} .$$

Since $\tilde{Q}$ is finite dimensional, it is compact if considered on $\mathcal{E}$ and on $\mathcal{B}$ as well. We may thus apply Theorem 3 in [Sch] and get

(8) $$\dim \mathcal{N}_{\mathcal{B}}(I + \tilde{Q}) = \dim \mathcal{N}_{\mathcal{E}'}(I' + \tilde{Q}') ,$$

(9) $$\mathcal{R}_{\mathcal{B}}(I + \tilde{Q}) = \{h \in \mathcal{B} ; \langle h, \nu \rangle = 0 \text{ for all } \nu \in \mathcal{N}_{\mathcal{E}'}(I' + \tilde{Q}')\} .$$

In view of the identity $I + \tilde{Q} = (I + Q_1)^{-1}K$ we have

(10) $$\mathcal{N}_{\mathcal{B}}(I + \tilde{Q}) = \mathcal{N}_{\mathcal{B}}(K) .$$

The identity $I' + \tilde{Q}' = K'(I' + Q_1')^{-1}$ shows that

$$(I' + Q_1')^{-1} : \mathcal{N}_{\mathcal{E}'}(I' + \tilde{Q}') \rightarrow \mathcal{N}_{\mathcal{E}'}(K')$$

is an isomorphism, so that

(11) $$\dim \mathcal{N}_{\mathcal{E}'}(I' + \tilde{Q}') = \dim \mathcal{N}_{\mathcal{E}'}(K') .$$

We observed in (6), (7) above that $g \in \mathcal{R}_{\mathcal{B}}(K)$ iff $h = (I + Q_1)^{-1}g \in \mathcal{R}_{\mathcal{B}}(I + \tilde{Q})$; since $\langle g, (I' + Q_1')^{-1}\nu \rangle = \langle h, \nu \rangle$, this together with (9) yields (5). Combining (11), (8), (10) we get (4). The rest is easy.

<u>Remark.</u> We shall now draw several geometric consequences on the structure of the boundary B from the assumption (1). These auxiliary results will be useful later in connection with the investigation of the equations (2_0), (3_0).

<u>5.3. Lemma.</u> Suppose that $F \subset R^m$ is closed, $r > 0$,

$z \in R^m$ and denote for $\theta \in \Gamma$ by $N_r^F(\theta, z)$ the total number (possibly zero or $+\infty$) of all points in

$$F \cap \{z + \rho\theta ;\ 0 < \rho < r\} .$$

Then the function

$$\theta \longmapsto N_r^F(\theta,\ z)$$

is Borel measurable on Γ and if we put

$$W_r^F(z) = \int_\Gamma N_r^F(\theta, z) d\mathcal{H}_{m-1}(\theta) ,$$

then for any $z \in R^m \setminus F$

$$W_r^F(z) \leqq [\mathrm{dist}(z, F)]^{1-m} \mathcal{H}_{m-1}(\Omega_r(z) \cap F) .$$

<u>Proof.</u> Fix an interval $\mathcal{J} = [a, b[$ with $0 < a < b < +\infty$ and denote for $\theta \in \Gamma$ by $N^{\mathcal{J}}(\theta, z)$ the number of points in

$$F \cap \{z + \rho\ \theta ;\ \rho \in \mathcal{J}\} .$$

For each positive integer n divide $\mathcal{J}$ into 2^n congruent intervals $I_1^n = [a,\ a + \frac{b-a}{2^n}[\ , \ldots , I_{2^n}^n = [b - \frac{b-a}{2^n}\ , b[$. It is easily seen that each of the sets

$$F_j = F \cap \{x \in R^m;\ |x-z| \in I_j^n\}$$

$(j = 1, \ldots, 2^n)$ is of the type F_σ . Consider now the map $R^m \setminus \{z\} \rightarrow \Gamma$ defined by

$$(12) \qquad \pi : x \longmapsto \frac{x-z}{|x-z|} , \quad x \in R^m \setminus \{z\} .$$

Clearly, all the sets $\pi(F_j)$ are of the type F_σ . Writing χ_M for the characteristic function of $M \subset \Gamma$ we conclude that

$$\theta \mapsto N_n(\theta) = \sum_{j=1}^{2^n} \chi_{\pi(F_j)}(\theta)$$

is Borel measurable on Γ. Since

$$y_1, y_2 \in F \setminus \Omega_a(z) \Longrightarrow |\pi(y_1) - \pi(y_2)| \leq a^{-1}|y_1-y_2|$$

we get

$$\mathcal{H}_{m-1}(\pi(F_j)) \leq a^{1-m}\, \mathcal{H}_{m-1}(F_j), \quad 1 \leq j \leq 2^n ,$$

whence

$$\int_{\Gamma} N_n(\theta) d\mathcal{H}_{m-1}(\theta) \leq a^{1-m}\, \mathcal{H}_{m-1}(F \cap \Omega_b(z)) .$$

It follows from our construction that $N_n(\theta)$ is the number of those intervals $I^n_1, \dots, I^n_{2^n}$ which contain at least one point of the form $|x-z|$ with $x \in \pi^{-1}(\theta)$. Hence we see that the sequence $\{N_n(\theta)\}_{n=1}^{\infty}$ tends non-decreasingly to $N^{\mathcal{J}}(\theta,z)$. Consequently, $\theta \mapsto N^{\mathcal{J}}(\theta,z)$ is Borel measurable and

$$(13) \qquad \int_{\Gamma} N^{\mathcal{J}}(\theta,z) d\mathcal{H}_{m-1}(\theta) \leq a^{1-m}\, \mathcal{H}_{m-1}(\Omega_b(z) \cap F) .$$

It remains to realize that $]0,r[$ may be expressed as a union of an increasing sequence of intervals $\mathcal{J}_n = [a_n,b_n[$; consequently, $N^F_r(\theta,z)$ - being a limit of the non-decreasing sequence $N^{\mathcal{J}_n}(\theta,z)$ - is Borel measurable. If $z \in R^m \setminus F$ and $a = \mathrm{dist}(z,F)$, then

$$N^F_r(\theta,z) = 0 \quad \text{for} \quad r \leq a$$

and

$$N_r^F(\theta, z) = N^{[a,r[}(\theta, z) \quad \text{for} \quad r > a .$$

This together with (13) completes the proof.

5.4. Lemma. If $H \subset R^m$ is a Borel set with a compact boundary ∂H such that $\mathcal{H}_{m-1}(\partial H) < +\infty$, then $P(H) < +\infty$ (cf. 2.10 for notation) and also each component of H has finite perimeter.

Proof. Fix $z^1, \ldots, z^{m+1} \in R^m \setminus \partial H$ in general position (i.e. not situated on a single hyperplane). According to 5.3 we have

$$W_\infty^{\partial H}(z^j) < +\infty$$

for each j ; this together with the inequalities

$$v^H(z^j) \leqq W_\infty^{\partial H}(z^j)$$

implies $P(H) < +\infty$ by 2.12. It remains to notice that, for each component H_o of H, $\partial H_o \subset \partial H$ and, consequently, $\mathcal{H}_{m-1}(\partial H_o) < +\infty$.

Remark. It can be shown that compactness of ∂H is irrelevant for validity of the above lemma.

5.5. Lemma. If G fulfils (1), then $\mathcal{H}_{m-1}(B) < +\infty$ and $\mathcal{H}_{m-1}(B \setminus \hat{B}) = 0$. If H is an arbitrary component of G, then

$$V_o^H \leqq V_o^G < \frac{1}{2}$$

and $n^H(y) = n^G(y) \neq 0$ for $\mathcal{H}_{m-1}$ - a.e. $y \in \partial H$ (cf. 2.14 for notation), $0 < \mathcal{H}_{m-1}(\partial H) < +\infty$.

<u>Proof.</u> We have observed in §4 that (1) implies the existence of an $\varepsilon \in]0,1[$ such that

(14) $\quad y \in B \Longrightarrow \varepsilon < d_G(y) < 1-\varepsilon .$

Combining this with the isoperimetric lemma stated in 2.14 we get

$$\liminf_{r \to 0+} r^{1-m} \mathcal{H}_{m-1}(\Omega_r(y) \cap \hat{B}) > 0$$

for all $y \in B$. Since the (m-1)-dimensional density of $\hat{B}$ equals zero $\mathcal{H}_{m-1}$ - a.e. in $B \setminus \hat{B}$ (cf. 2.10 in [Fe 4]), we conclude that

(15) $$\mathcal{H}_{m-1}(B \setminus \hat{B}) = 0 ,$$

so that $0 < \mathcal{H}_{m-1}(B) = \mathcal{H}_{m-1}(\hat{B}) < +\infty$ (cf. 2.14). According to (15), $n^G(y)$ is well-defined and different from the zero vector for $\mathcal{H}_{m-1}$ - a.e. $y \in B$. Consider now an arbitrary component H of G . We know from 5.4 that necessarily $P(H) < < +\infty$. Fix an arbitrary $y \in \hat{B} \cap \widehat{\partial H}$. It follows from the inclusion $H \subset G$ and the relation

$$\lim_{r \to 0+} r^{-m} \mathcal{H}_m(\{x \in \Omega_r(y) \setminus H;\ (x-y)n^H(y) > 0\}) = 0$$

that also

$$\lim_{r \to 0+} r^{-m} \mathcal{H}_m(\{x \in \Omega_r(y) \setminus G;\ (x-y)n^H(y) > 0\}) = 0 ;$$

this together with the existence of a non-zero $n^G(y)$ yields $n^H(y) = n^G(y)$. In view of (15) we have thus

$$n^H(.) = n^G(.)\ \mathcal{H}_{m-1} \text{ - a.e. on } \widehat{\partial H} .$$

Hence we get by results in 2.15 and in §1 that

$$v_r^H(z) = |\lambda_z^H|(\Omega_r(z) \setminus \{z\}) =$$

$$= \int_{\hat{\partial} H \cap \Omega_r(z)} |n^H(y) \cdot \operatorname{grad} h_z(y)| \, d\mathcal{H}_{m-1}(y) =$$

$$= \int_{\hat{\partial} H \cap \Omega_r(z)} |n^G(y) \cdot \operatorname{grad} h_z(y)| \, d\mathcal{H}_{m-1}(y) \leqq$$

$$\leqq \int_{B \cap \Omega_r(z)} |n^G(y) \cdot \operatorname{grad} h_z(y)| \, d\mathcal{H}_{m-1}(y) = v_r^G(z) ,$$

so that

$$V_o^H = \lim_{r \to 0+} \sup_{z \in \partial H} v_r^H(z) \leqq \lim_{r \to 0+} \sup_{z \in B} v_r^G(z) = V_o^G < \frac{1}{2} .$$

By the first part of the proof we conclude that

$$\mathcal{H}_{m-1}(\partial H \setminus \hat{\partial} H) = 0 , \quad 0 < \mathcal{H}_{m-1}(\partial H) = \mathcal{H}_{m-1}(\hat{\partial} H) < +\infty$$

and the proof is complete.

5.6. Lemma. If G fulfils (1), then the number of its bounded components does not exceed $\dim \mathcal{N}_{\mathcal{B}}(W)$. If H is a bounded component of G and f_H is defined by (16), then $f_H \in \mathcal{N}_{\mathcal{B}}(W)$.

Proof. Consider an arbitrary bounded component H of G and define f_H on B as follows:

$$(16) \qquad f_H(y) = \frac{d_H(y)}{d_G(y)} , \quad y \in B .$$

Since the density of any Borel set is of the first class of Baire and (1) implies (14), we conclude that $f_H \in \mathcal{B}$. We shall now calculate $Wf_H(y) = \langle f_H, \lambda_y^G \rangle$ for $y \in B$. If $y \in B \setminus \partial H = B \setminus \mathrm{cl}\, H$, then $f_H(y) = 0$ and noting that, in view of 5.5,

$$(16_1) \qquad f_H = 1 \quad \mathcal{H}_{m-1}\text{ - a.e. on } \partial H ,$$

we get by 2.15 and 2.8

$$\langle f_H, \lambda_y^G \rangle = \int_{B \setminus \{y\}} f_H d\, \lambda_y^G =$$

$$= - \int_{\partial H} f_H(x) n^G(x) \cdot \mathrm{grad}\, h_y(x) d\, \mathcal{H}_{m-1}(x) =$$

$$= - \int_{\partial H} n^H(x) \cdot \mathrm{grad}\, h_y(x) d\, \mathcal{H}_{m-1}(x) = \lambda_y^H(\partial H \setminus \{y\}) =$$

$$= - d_H(y) = 0 .$$

If $y \in \partial H$, then 2.8, 2.15 yield

$$\langle f_H, \lambda_y^G \rangle = f_H(y) d_G(y) -$$

$$- \int_{\partial H \setminus \{y\}} f_H(x) n^G(x)\, \mathrm{grad}\, h_y(x) d\, \mathcal{H}_{m-1}(x) =$$

$$= d_H(y) - \int_{\partial H \setminus \{y\}} n^H(x) \cdot \mathrm{grad}\, h_y(x) d\, \mathcal{H}_{m-1}(x) =$$

$$= d_H(y) - \lambda_y^H(\partial H \setminus \{y\}) = 0 .$$

We see that $f_H \in \mathcal{N}_{\mathcal{B}}(W)$. Let now $H_1, \ldots, H_q$ be different bounded components of G. For $i \neq j$ we have

$$n^{H_i}(.) \neq n^{H_j}(.) \quad \text{on} \quad \widehat{\partial} H_i \cap \widehat{\partial} H_j \ ;$$

on the other hand, we conclude from lemma 5.5 that $n^{H_i}(.) =$ $= n^G(.)$ a.e. ($\mathcal{H}_{m-1}$) on ∂H_i and, consequently,

$$\mathcal{H}_{m-1}(\partial H_i \cap \partial H_j) = 0 \ .$$

Noting that $\mathcal{H}_{m-1}(\partial H_i) > 0$ and using (16) we see that $f_{H_1}, \dots, f_{H_q}$ are linearly independent, so that $q \leq \mathcal{N}_{\mathcal{B}}(W)$.

<u>5.7. Remark.</u> Let us recall that the potential $\mathcal{U}\nu$ of any signed measure ν with a compact support in R^m is defined a.e. ($\mathcal{H}_m$) and is locally integrable; besides that, for each $\varphi \in \mathcal{D}$

$$\int_{R^m} \varphi \, d\nu = - \int_{R^m} \Delta \varphi(x) . \mathcal{U}\nu(x) dx$$

(which means that, in the language of distribution theory, $\Delta \mathcal{U}\nu = -\nu$). Indeed, we observed in §1 that this is true for $\nu = \delta_y$ (= the Dirac measure at y). The general case reduces to this special case on account of the identity

$$\mathcal{U}\nu(x) = \int_{R^m} \mathcal{U}\delta_y(x) d\nu(y)$$

with help of Fubini's theorem.

<u>5.8. Lemma.</u> Let $\mathcal{H}_m(B) = 0$. If $\mu \in \mathcal{C}'_c(B)$ (cf. 2.21 for notation), then

$$(17) \qquad \int_{R^m} |\operatorname{grad} \mathcal{U}\mu(x)|^2 dx = \int_{R^m} \mathcal{U}_c \mu \, d\mu .$$

If $\nu \in \mathcal{N}_{\mathcal{C}'}(W')$, then $\mathcal{U}\nu$ remains constant on each component of int G .

<u>Proof.</u> We know that $\mathcal{U}\mu$ is infinitely differentiable in $R^m \setminus B$ and there is a continuous function $\mathcal{U}_c \mu$ on R^m such that $\mathcal{U}\mu = \mathcal{U}_c \mu$ in $R^m \setminus B$ (i.e.a.e.). Fix a symmetric function $\omega_n \geqq 0$ in $\mathfrak{D}$ such that

$$\int_{R^m} \omega_n(x) dx = 1 , \quad \operatorname{spt} \omega_n \subset \Omega_{\frac{1}{n}}(0)$$

and consider the function

$$\mathcal{U}(\mu * \omega_n) = (\mathcal{U}\mu) * \omega_n$$

which is infinitely differentiable on R^m (cf. remarks following 2.22). For every positive integer k construct a $\psi_k \in \mathfrak{D}$ such that

$$0 \leqq \psi_k \leqq 1 , \quad \psi_k(\Omega_k(0)) = \{1\} , \quad |\operatorname{grad} \psi_k| \leqq 1 .$$

We have then for large k (recall that $\mu * \omega_n$ has a compact support)

$$\int_{R^m} \mathcal{U}(\mu * \omega_n) d(\mu * \omega_n) =$$

$$= \int_{R^m} \psi_k \mathcal{U}(\mu * \omega_n) d(\mu * \omega_n) =$$

$$= - \int_{R^m} \Delta\,[\psi_k\,\mathcal{U}(\mu * \omega_n)]\,\mathcal{U}(\mu * \omega_n)\,d\mathcal{H}_m =$$

$$= \int_{R^m} \operatorname{grad}\,[\psi_k\,\mathcal{U}(\mu * \omega_n)]\cdot\operatorname{grad}\,\mathcal{U}(\mu * \omega_n)\,d\mathcal{H}_m =$$

$$= \int_{R^m} (\operatorname{grad}\,\psi_k)\cdot\mathcal{U}(\mu * \omega_n)\,\operatorname{grad}\,\mathcal{U}(\mu * \omega_n)\,d\mathcal{H}_m +$$

$$+ \int_{R^m} \psi_k\,|\operatorname{grad}\,\mathcal{U}(\mu * \omega_n)|^2\,d\mathcal{H}_m .$$

Note that $|\operatorname{grad}\,\psi_k| = 0$ on $\Omega_k(0)$, $|\operatorname{grad}\,\psi_k| \leqq 1$ and, for fixed n , $\mathcal{U}(\mu * \omega_n)$ is bounded and $|\operatorname{grad}\,\mathcal{U}(\mu * \omega_n)|$ is integrable. Consequently,

$$\lim_{k\to\infty} \int_{R^m} \operatorname{grad}\,\psi_k\,\mathcal{U}(\mu * \omega_n)\,\operatorname{grad}\,\mathcal{U}(\mu * \omega_n)\,d\mathcal{H}_m = 0 .$$

Clearly,

$$\lim_{k\to\infty} \int_{R^m} \psi_k|\operatorname{grad}\,\mathcal{U}(\mu * \omega_n)|^2\,d\mathcal{H}_m =$$

$$= \int_{R^m} |\operatorname{grad}\,\mathcal{U}(\mu * \omega_n)|^2\,d\mathcal{H}_m .$$

Since $\mathcal{U}(\mu * \omega_n) = (\mathcal{U}_c\,\mu) * \omega_n \to \mathcal{U}_c\mu$ locally uniformly and $\mu * \omega_n \to \mu$ vaguely as $n \to \infty$ (cf. remarks following 2.22 and note that the supports of all $\mu * \omega_n$ are contained in a fixed compact set) we get

$$\lim_{n\to\infty} \int_{R^m} \mathcal{U}(\mu * \omega_n) d(\mu * \omega_n) = \int_{R^m} \mathcal{U}_c \mu \, d\mu ,$$

so that

$$(18) \qquad \int \mathcal{U}_c \mu \, d\mu = \lim_{n\to\infty} \int_{R^m} |\operatorname{grad} \mathcal{U}(\mu * \omega_n)|^2 \, d\mathcal{H}_m .$$

Making use of Fatou's lemma and the fact that $\mathcal{U}\mu$ is continuously differentiable in $R^m \setminus B$ so that

$\lim_{n\to\infty} |\operatorname{grad} \mathcal{U}(\mu * \omega_n)| = |\operatorname{grad} \mathcal{U}\mu|$ in $R^m \setminus B$ (i.e.a.e.)

we get finally

$$(19) \qquad \int_{R^m} |\operatorname{grad} \mathcal{U}\mu|^2 \, d\mathcal{H}_m \leqq \int \mathcal{U}_c \mu \, d\mu .$$

Notice that $\mathcal{U}(\mu * \omega_n) - \mathcal{U}\mu = \mathcal{U}(\mu * \omega_n - \mu)$ is infinitely differentiable in $R^m \setminus B$ and coincides there with $\mathcal{U}(\mu * \omega_n) - \mathcal{U}_c \mu$ which is a function continuous on the whole R^m. The above reasoning which led to (19) may thus be repeated with μ replaced by $\mu * \omega_n - \mu$ to yield

$$\int_{R^m} |\operatorname{grad} \mathcal{U}(\mu * \omega_n) - \operatorname{grad} \mathcal{U}\mu|^2 d\mathcal{H}_m =$$

$$= \int_{R^m} |\operatorname{grad} \mathcal{U}(\mu * \omega_n - \mu)|^2 d\mathcal{H}_m \leqq$$

$$\leqq \int_{R^m} [\mathcal{U}(\mu * \omega_n) - \mathcal{U}_c \mu] \, d(\mu * \omega_n - \mu) .$$

Since $\mathcal{U}(\mu * \omega_n) - \mathcal{U}_c\mu = (\mathcal{U}_c\mu) * \omega_n - \mathcal{U}_c\mu \to 0$ locally uniformly and $\mu * \omega_n - \mu \to 0 \in \mathcal{E}'$ vaguely as $n \to \infty$, we arrive at

$$(20)\quad \lim_{n\to\infty} \int_{R^m} |\text{grad } \mathcal{U}(\mu * \omega_n) - \text{grad } \mathcal{U}\mu|^2 d\,\mathcal{H}_m = 0\,.$$

In particular,

$$\lim_{n\to\infty} \int_{R^m} |\text{grad } \mathcal{U}(\mu * \omega_n)|^2 d\,\mathcal{H}_m = \int_{R^m} |\text{grad } \mathcal{U}\mu|^2 d\,\mathcal{H}_m$$

which together with (18) proves (17).

Suppose now that $\mu \in \mathcal{N}_{\mathcal{E}'}(W')$. Since $\mathcal{H}_m(B) = 0$ and the operator $W = W_B^G$ remains unchanged if G is replaced by an equivalent set (cf. remark 2.3), we may assume that $G = G \setminus B$ is open, so that

$$W' = N^G \mathcal{U}$$

by 2.20. We know from §4 that necessarily $\mu \in \mathcal{E}_c'(B)$. In view of $\psi_k \cdot \mathcal{U}(\mu * \omega_n) \in \mathcal{D}$ we have

$$0 = \langle \psi_k \cdot \mathcal{U}(\mu * \omega_n), N^G \mathcal{U}\mu \rangle =$$

$$= \int_G (\text{grad } \psi_k) \cdot \mathcal{U}(\mu * \omega_n) \cdot \text{grad } \mathcal{U}\mu \; d\,\mathcal{H}_m +$$

$$+ \int_G \psi_k [\text{grad } \mathcal{U}(\mu * \omega_n)] \text{ grad } \mathcal{U}\mu \; d\,\mathcal{H}_m\,.$$

Making first $k \to \infty$ and noting that

$$\left|\int_G (\operatorname{grad} \psi_k)\cdot \mathcal{U}(\mu * \omega_n)\cdot \operatorname{grad} \mathcal{U}\mu \, d\mathcal{H}_m\right| \leq$$

$$\leq \int_{R^m \setminus \Omega_k(0)} |\mathcal{U}(\mu * \omega_n)|\cdot|\operatorname{grad} \mathcal{U}\mu| \, d\mathcal{H}_m \to 0 ,$$

$$\lim_{k\to\infty} \int_G \psi_k [\operatorname{grad} \mathcal{U}(\mu * \omega_n)]\cdot \operatorname{grad} \mathcal{U}\mu \, d\mathcal{H}_m =$$

$$= \int_G [\operatorname{grad} \mathcal{U}(\mu * \omega_n)]\cdot \operatorname{grad} \mathcal{U}\mu \, d\mathcal{H}_m$$

we get

$$0 = \int_G [\operatorname{grad} \mathcal{U}(\mu * \omega_n)]\cdot \operatorname{grad} \mathcal{U}\mu \, d\mathcal{H}_m$$

for all n . Hölder's inequality combined with (20) gives

$$\int_G |\operatorname{grad} \mathcal{U}\mu|^2 d\mathcal{H}_m =$$

$$= \lim_{n\to\infty} \int_G [\operatorname{grad} \mathcal{U}(\mu * \omega_n)]\cdot \operatorname{grad} \mathcal{U}\mu \, d\mathcal{H}_m = 0 ,$$

so that $\mathcal{U}\mu$ must remain constant on each component of int $G = G$.

5.9. Proposition. Assume (1). Then $\partial(\text{int } G) = \partial(\text{int } C) = B$, G has a finite number of components and their closures are mutually disjoint. If $G_1,\ldots,G_p$ are all bounded components of G and χ_j denotes the characteristic function of ∂G_j on B $(j = 1,\ldots,p)$, then $\{\chi_1,\ldots,\chi_p\}$

is a basis in

$$\mathcal{N}_{\mathcal{C}}(W) = \mathcal{N}_{\mathcal{B}}(W) .$$

5.10. Remark. We shall prove 5.9 together with proposition 5.11 below. For its formulation we adopt the following terminology. If G satisfies (1) and there exists a $\mu \in \mathcal{C}'(B)$ with $\mu(B) \neq 0$ such that $\mathcal{U}\mu$ vanishes identically on $\operatorname{int} G$, then G will be called critical. (We shall see below that this can happen only if G is a bounded set in the plane.) An example of a critical set is provided by the unit disc in R^2. Indeed, if $G = \Omega_r(0) \subset R^2$ and $\mu \in \mathcal{C}'(\partial \Omega_r(0))$ has a constant density with respect to $\mathcal{H}_1$ on $\partial \Omega_r(0)$, then elementary calculation shows that

$$(21) \quad |x| \leqq r \Longrightarrow \mathcal{U}\mu(x) = \frac{1}{2\pi} \mu(\partial \Omega_r(0)) \log \frac{1}{r}$$

(cf. [KNV III]).

5.11. Proposition. Assume (1). Then $\dim \mathcal{N}_{\mathcal{C}'}(W')$ equals the number of bounded components of G. Let G_j have the same meaning as in 5.9. If G is not critical, then there is a $\mu_j \in \mathcal{C}'(B)$ such that $\mathcal{U}\mu_j = 1$ on $\operatorname{int} G_j$ and $\mathcal{U}\mu_j = 0$ on $\operatorname{int}(G \setminus G_j)$ $(j = 1,\ldots,p)$; $\{\mu_1,\ldots,\mu_p\}$ is a basis in $\mathcal{N}_{\mathcal{C}'}(W')$. If G is critical (which can occur only if $m = 2$ and G is bounded), we can fix $\mu_1 \in \mathcal{C}'(B)$ with $\mu_1(B) \neq 0$ having identically vanishing potential on $\operatorname{int} G$; further choose $x_j \in \operatorname{int} G_j$ $(j = 1,\ldots,p)$. There exist $\mu_2,\ldots,\mu_p \in \mathcal{C}'$ with $\mu_k(B) = 0$ having constant $\mathcal{U}\mu_k$ on each of the sets $\operatorname{int} G_j$ such that the matrix $\mathcal{U}\mu_k(x_j)$

$(2 \leq k \leq p,\ \ 1 \leq j \leq p)$ has rank $p-1$. Then $\{\mu_1, \mu_2, \dots, \mu_p\}$ is a basis in $\mathcal{N}_{\mathcal{C}'}(W')$ and the mapping $\mathcal{N}_{\mathcal{C}'}(W') \to R^p$ given by

$$(22) \qquad \mu \longmapsto (\mathcal{U}\mu(x_1), \dots, \mathcal{U}\mu(x_p))$$

does not contain the vector $(1, \dots, 1) \in R^p$ in its range.

<u>Proof.</u> We observed in remark 2.3 that the operator $W = W_B^G$ is not affected by changes of G in $\mathcal{H}_m$-measure zero. Since $\mathcal{H}_m(B) = 0$, we may therefore assume that $G = G \setminus B$ is open, so that $W' = N^G \mathcal{U}$ by 2.20. If $\nu \in \mathcal{N}_{\mathcal{C}'}(W')$ then $\mathcal{U}\nu$ remains constant on each component of G by 5.8. If G is unbounded and G_o is the unbounded component of G, then $\mathcal{U}\nu$ must vanish on G_o. This is clear in case $m > 2$, because then $\mathcal{U}\nu$ tends to zero at infinity, while for $m = 2$ the relation

$$\lim_{|x| \to \infty} \left| \mathcal{U}\nu(x) + \frac{1}{2\pi} \nu(B) \log |x| \right| = 0$$

shows that the potential $\mathcal{U}\nu$ can remain constant on G_o only if $\nu(B) = 0$ when its limit at infinity equals zero. Consider first the case when G is not critical. We shall show that the mapping (22) is a monomorphism on $\mathcal{N}_{\mathcal{C}'}(W')$. Indeed, if $\nu \in \mathcal{N}_{\mathcal{C}'}(W')$ and $\mathcal{U}\nu$ vanishes identically on $G_1 \cup \dots \cup G_p$, then $\mathcal{U}\nu$ tends to zero at infinity (because now we are not in the critical case) and, in view of $\nu \in \mathcal{E}_c'(B)$, $\mathcal{U}\nu$ has zero limits at B, so that $\mathcal{U}\nu = 0$ on $R^m \setminus B$. Noting that $\mathcal{H}_m(B) = 0$ and using remark 5.7 we get

$$\langle \varphi, \nu \rangle = - \int_{R^m} \Delta \varphi\, \mathcal{U}\nu \, d\mathcal{H}_m = 0 \,, \qquad \varphi \in \mathcal{D} \,,$$

so that ν must be trivial. Hence $\dim \mathcal{N}_{\mathcal{C}'}(W') \leqq p$. This together with lemma 5.2 and the inequality $p \leq \mathcal{N}_{\mathcal{B}}(W)$ established in 5.6 gives

$$(23) \qquad \dim \mathcal{N}_{\mathcal{B}}(W) = \dim \mathcal{N}_{\mathcal{C}'}(W') = p \,.$$

In particular, (22) is an isomorphism between $\mathcal{N}_{\mathcal{C}'}(W')$ and R^p and for each j we can choose $\mu_j \in \mathcal{N}_{\mathcal{C}'}(W')$ which is mapped by (22) onto the vector $(\delta_{j1}, \ldots, \delta_{jp}) \in R^p$ ($j = 1, \ldots, p$, δ_{jk} is Kronecker's symbol); $\{\mu_1, \ldots, \mu_p\}$ is a basis in $\mathcal{N}_{\mathcal{C}'}(W')$. Since $\mu_j \in \mathcal{E}'_c(B)$ we see that necessarily $\mathrm{cl}\, G_j \cap \mathrm{cl}\, G_k = \emptyset$ for $j \neq k$ (because $\mathcal{U}\mu_j = 1$ on G_j and $\mathcal{U}\mu_j = 0$ on G_k) and, for the same reason, $\mathrm{cl}\, G_j \cap \mathrm{cl}\, G_o = \emptyset$ if G is unbounded and G_o is its unbounded component ($j = 1, \ldots, p$) .

Now consider the case when G is critical and fix $\mu_1 \in \mathcal{C}'$ with vanishing $\mathcal{U}\mu_1$ on G and $\mu_1(B) \neq 0$. The above reasoning shows that the mapping (22), if restricted to

$$\mathcal{N}_o = \{\mu \in \mathcal{N}_{\mathcal{C}'}(W');\ \mu(B) = 0\} \,,$$

is a monomorphism $\mathcal{N}_o \to R^p$. It is easily seen that $\{\mu_1\} \cup \mathcal{N}_o$ spans the whole $\mathcal{N}_{\mathcal{C}'}(W')$. We shall show that the range of the map (22) on $\mathcal{N}_o$ omits the vector $(1, \ldots, 1) \in R^p$. Admitting the contrary we fix a $\varkappa \in \mathcal{N}_o$ with $\mathcal{U}\varkappa = 1$ on G (note that G , being critical, must be bounded). Fix $r > 1$ large enough to guarantee

(24) $$\operatorname{cl} G \subset \Omega_r(0)$$

and consider a probability measure μ distributed on $\partial\Omega_r(0)$ with a constant density with respect to $\mathcal{H}_1$. As noticed in (21) in 5.10,

$$\mathcal{U}\mu = \frac{1}{2\pi} \log \frac{1}{r} \quad \text{on} \quad \Omega_r(0) \supset \operatorname{cl} G \; .$$

Fubini's theorem implies the reciprocity law

(25) $$\int_{R^2} \mathcal{U}\varkappa \, d\mu = \int_{R^2} \mathcal{U}\mu \, d\varkappa \; .$$

Now $\mathcal{U}\varkappa$ (being harmonic on $R^2 \setminus \operatorname{cl} G$ and tending to 1 at $\partial(R^2 \setminus \operatorname{cl} G)$ and to zero at infinity) remains positive on $R^2 \setminus \operatorname{cl} G \supset \partial\Omega_r(0)$, so that the left-hand side of (25) is positive, while the right-hand side equals $\varkappa(B) \cdot \frac{1}{2\pi} \log \frac{1}{r} = 0$. This contradiction shows that $\dim \mathcal{N}_0 \leqq p-1$, whence $\dim \mathcal{N}_{\mathcal{C}'}(W') \leqq p$ also in the critical case and (23) is established in full generality. If $G \subset R^2$ is any bounded set and (24) holds, then $\tilde{G} = G \cup (R^2 \setminus \operatorname{cl} \Omega_r(0))$, being unbounded, cannot be critical. Consequently, its components must have disjoint closures.

We see that any G fulfilling (1) has components with disjoint closures. In particular, the functions $\varkappa_j$ defined in 5.9 belong to $\mathcal{C}$ and, on account of 5.6, also to $\mathcal{N}_{\mathcal{B}}(W)$ $(j = 1,\dots,p)$; being linearly independent they form a basis in $\mathcal{N}_{\mathcal{B}}(W)$ and $\mathcal{N}_{\mathcal{B}}(W) = \mathcal{N}_{\mathcal{C}}(W)$. Thus the proof of propositions 5.9 and 5.11 is complete.

Remark. The above propositions 5.9 and 5.11 describing null-spaces of the operators $W = \frac{1}{2}(I + T)$, $W' = \frac{1}{2}(I' + T')$ combined with the Fredholm alternative yield existence theorems on boundary value problems.

5.12. Theorem on the Neumann problem. Let $G \subset R^m$ be an open set with a compact boundary satisfying (1) and consider the Neumann problem

$$N^G \mathcal{U}\nu = \mu \tag{26}$$

with a prescribed $\mu \in \mathcal{C}'(B)$. Then (26) admits a solution $\nu \in \mathcal{C}'(B)$ iff $\mu(\partial H) = 0$ for each bounded component H of G . The solution ν is uniquely determined iff G is unbounded and connected.

Proof. We know that (26) is equivalent to (3) with $\mu_1 = 2\mu$. According to (1) we have $\omega T < 1$. It remains to apply 5.2 and 5.9 together with 5.11.

5.13. Theorem on the Dirichlet problem. Suppose that $C \subset R^m$ satisfies $V_o^C < \frac{1}{2}$ and let $G_1, \ldots, G_p$ be bounded components of $G = R^m \setminus C$; then $\operatorname{int} G_j \neq \emptyset$ and we may fix $x_j \in \operatorname{int} G_j$ $(j = 1, \ldots, p)$. Given $g \in \mathcal{C}(\partial C)$, then there are constants $c_o, c_1, \ldots, c_p$ and an $f \in \mathcal{C}(\partial C)$ such that

$$W^G f(.) + c_o + \sum_{j=1}^{p} c_j h_{x_j}(.) \tag{27}$$

represents a solution of the Dirichlet problem for $\operatorname{int} C$ and the boundary condition g . If G is not critical, we may put

$c_o = 0$ and the constants $c_1,\dots,c_p$ are then uniquely determined; with this choice of c_o, f is uniquely determined iff C is bounded and G connected (in which case the sum $\sum_{j=1}^{p} c_j h_{x_j}(\cdot)$ in (27) disappears). If G is critical (which can occur only if $m = 2$ and G is bounded), then c_o is uniquely determined and $c_1,\dots,c_p$ are uniquely determined by the additional requirement

$$c_1 + \dots + c_p = 0 .$$

<u>Proof.</u> Recall that $V_o^C = V_o^G$, $W = \frac{1}{2}(I + T)$, $\omega T = 2V_o^G < 1$. Let μ_j, x_j $(j = 1,\dots,p)$ have the meaning described in 5.11 and put $g_1 = 2g$. Consider first the case when G is not critical. According to 5.2, the function

$$\breve{g} = g_1 - \sum_{j=1}^{p} c_j h_{x_j}$$

will belong to $\mathcal{R}_{\mathcal{C}}(I + T)$ iff

$$\langle \breve{g}, \mu_j \rangle = 0 , \quad 1 \leqq j \leqq p ;$$

from the construction of the basis we get

$$c_j = \langle g_1, \mu_j \rangle , \quad j = 1,\dots,p .$$

Next suppose that G is critical. Now the function

$$g^* = g_1 - c_o - \sum_{j=1}^{p} c_j h_{x_j}$$

can be in $\mathcal{R}_{\mathcal{C}}(I + T)$ only if $0 = \langle g^*, \mu_1 \rangle$ which means (note that $\mathcal{U}\mu_1$ vanishes on $\operatorname{int} G$, so that $\langle h_{x_j}, \mu_1 \rangle = 0$

for $j = 1,\dots,p$) that

$$c_0 = \langle g_1, \mu_1 \rangle .$$

Defining c_0 by this equality we see that $g^* \in \mathcal{R}_{\mathcal{C}}(W)$ iff

$$(28) \qquad \langle g^*, \mu_k \rangle = 0 \quad \text{for} \quad k = 2,\dots,p .$$

Since $\mu_k(B) = 0$ $(k = 2,\dots,p)$ we observe that (28) is fulfilled iff

$$(29) \qquad \sum_{j=1}^{p} c_j \,\mathcal{U}\mu_k(x_j) = \langle g_1, \mu_k \rangle, \quad 2 \leqq k \leqq p .$$

The matrix $\mathcal{U}\mu_k(x_j)$ $(1 \leqq j \leqq p,\ 2 \leqq k \leqq p)$ has rank $p-1$ and we know from 5.11 that the vector $(1,\dots,1) \in R^p$ cannot be expressed as a linear combination of the vectors $(\mathcal{U}\mu_k(x_1),\dots,\mathcal{U}\mu_k(x_p)) \in R^p$, $k = 2,\dots,p$. Consequently, $c_1,\dots,c_p$ will be uniquely determined if we adjoin the equation

$$c_1 + \dots + c_p = 0$$

to the system (29). We know from 5.9 that f will be uniquely determined only if $p = 0$ (when the sum $\sum_{j=1}^{p} c_j h_{x_j}$ disappears); in this case G is not critical and the choice $c_0 = 0$ determines f uniquely.

5.14. Remark. In the above theorem we treated the Dirichlet problem on account of the equation (2) considered over $\mathcal{C}$. The alternative 5.2 and propositions 5.9, 5.11 permit to investigate the same equation over $\mathcal{B}$ as well. This would lead to results on integral representability of generalized solutions of the Dirichlet problem where the boundary condition may be discontinuous. We shall not treat this topic here.

§ 6

Comments and references

In this paragraph we collect comments and references that were omitted through the preceding text which represents an expanded version of lectures delivered in a seminar held at Instituto de Matemática, Universidade Estadual de Campinas (S. P., Brasil) in 1978. Some parts of the material included here were presented in a seminar held at Brown University (R. I., U.S.A.) in 1965 and in various colloquium lectures during 1965 - 1978. The central theme is Fredholm's method for solving boundary value problems in potential theory without imposing à priori smoothness restrictions on the boundary of the domain. Although historical roots of the Fredholm theory of equations of the second kind lie in potential theoretic problems, most books on potential theory nowadays usually omit this topic which is, as a rule, included in courses of integral equations and partial differential equations of mathematical physics where, however, there is rarely adequate space for general investigation of boundary conditions for non-smooth boundaries. This gap in the literature was felt in potential theoretic courses held at Charles University in Prague (Czechoslovakia) during the last two decades and led to investigation of boundary behavior of potentials on general domains. The corresponding investigations of the logarithmic potentials (compare [K1]) and their application to boundary value problems in R^2 were included in the lecture notes [K I], [KNV II]. Here we present a unified treatment of logarithmic and Newtonian potentials in R^m for arbitrary $m \geq 2$ in the spirit of the papers [K 2], [K 8], [BM], [KN 1].

Introductory remark We suppose that the reader is acquainted with basic properties of harmonic functions (cf. [B], [C], [H], [HK], [La], [O 1], [dP], [W]). Classical potential theory with its sources is treated in [Ke], a presentation of potential theory oriented towards applications may be found in [Ma]. Investigation of potentials induced by charges distributed on smooth surfaces satisfying the so-called Ljapunov condition and their applications is presented in [G]. Applications of integral equations to problems concerning logarithmic potentials in the plane are described in [Lo], [RS], [D]. Integral equations arising in potential theory are also examined in [J], [Mi] (cf. [Lon] for a survey of various applications of integral equations). The reader interested in connections with the theory of partial differential equations is referred to [L]. A systematic presentation of the theory of elliptic partial differential equations may be found in [Mi C], potential theoretic methods for higher order partial differential equations are treated in [SW]. We do not attempt to give complete bibliography here. The interested reader will find further references in the above mentioned sources (cf. also [ASW], [KNV II], [KNV IV]).

§ 1 Weak characterization of the normal (or conormal) derivative is now standard in the theory of boundary value problems for partial differential equations. Already Plemelj, who introduced the concept of the so-called "Randströmung", was aware of the fact that a weak characterization of the Neumann boundary condition might be physically better intuitive than the pointwise normal derivative (cf. [Pl]). Functionals suitable for weak characterization of boundary values are investigated in [You]. The reader interested in concepts of distribution theory may consult [S].

The definition of a hit introduced in 1.7 generalizes that adopted in section 1.5 in [K 2] whose presentation we follow quite closely in §1; also some parts of the subsequent paragraphs are contained in [K 2]. Similar treatment of potential theoretic problems appeared in [BM].

The quantity $v_r^Q(.)$ defined in 1.11 proved to be useful in various investigations and we use it systematically in §2 in connection with the problem of continuous extendability of double layer potentials. It was generalized to higher codimensions [Br 1] and applied in geometric investigations [Br 2], [Br 3]. In plane a similar quantity (called cyclic variation) defined for arcs was used in [FK] to get estimates of the analytic capacity. Various modifications counting hits with suitable weight were employed in [Lu], [KL 1], [KL 2], [K 3]. Variants of this quantity where half-lines are replaced by parabolas were used for investigation of heat potentials (cf. [K 4], [Do 2], [Do 3], [Do 4], [V 1]). Comparison of the cyclic variation with a more restrictive concept called bend or rotation (= "Drehung") introduced by

Radon in [R 2] for the case when ∂Q is a Jordan curve in the plane is carried out in [ŠV].

[§ 2] Usually the double layer potentials are studied only on surfaces satisfying à priori smoothness restrictions like Ljapunov condition (cf. [G]). An important contribution to the theory of double layer logarithmic potentials was due to J. Radon [R 2] who treated boundary value problems on plane domains bounded by curves of bounded rotation (= "Kurven beschränkter Drehung"). There was a belief that such curves represent the utmost limit in generality for investigation of double layer potentials (cf. [RS]). As far as we know no direct generalization of Radon's boundaries was studied in higher dimensional spaces. Sections 2.5, 2.19, 2.20 show that the condition (37) occurring in 2.19 represents a natural necessary and sufficient restriction guaranteeing existence and continuous extendability of double layer potentials with continuous densities on the boundary ∂G of $G \subset R^m$. The quantity $v^G(.)$ permits also to derive simple necessary and sufficient conditions for the existence of angular limits of double layer potentials (cf. [Do 1], [K 5]).

Various properties of sets $G \subset R^m$ having finite perimeter $P(G)$ in the sense of 2.10 were investigated in [DG 1], [DG 2], [M], [Mi M], [Fe 2].

The symmetry rule presented in 2.23 was first discovered by Plemelj for plane domains bounded by regular curves (cf. [Pl]); for its general form in m-space cf. [BM].

§ 3 The method of successive approximations (which was used earlier by Liouville in another context - cf. [Kl]) was applied in 1845 to the Dirichlet problem in the plane by A. Beer who developed the solution in a formal series of logarithmic potentials (cf. [Be], [BMe]). The attempt to justify the convergence of the series obtained from the equation (4) in §3 led C. Neumann to his investigation [N 1] - [N 3] of contractivity (for convex domains) of the operator T called by him the operator of the arithmetical mean. G. Robin used similar arguments in connection with investigation of the equilibrium distribution of electricity on the surface of a conductor [Rob]. C. Neumann mentioned that generalization of the method of the arithmetic mean to spaces of higher dimensions was subject of a thesis by E. Riquier presented in Paris in 1886. (Riquier's thesis was not available to the present author.) Neumann's method led to further investigations of nonconvex domains by H. Poincaré, A. Korn, A. Ljapunov, J. Plemelj, V. Steklov, S. Zaremba and others (cf. [NE], [Li]). The role played by Neumann's method in the context of the mathematics in the 19 th century is described in [Kle], [So] (cf. also [HT]). In the later development Neumann's lemma on contractivity of the operator of the arithmetical mean retained its significance in connection with approximate solution to boundary value problems for harmonic functions (cf. [KK], [Gou]), constructing the Riemann mapping function for convex domains (cf. [Ga]) and related topics. C. Neumann's original proof [N 1] of his lemma contained a gap; that is why the lemma is often presented under additional stronger restrictions on the boundary (cf. [Scho] for a corrected treatment containing some comments and references).Neumann's error was sharply criticized by Lebesgue [Le 1] (reprinted in [Le 2], pp. 107-122) who gave a correct proof of the convergence of the Neumann series for arbitrary convex domains in the plane. Lebesgue was apparently unaware of Neumann's corrected treatment [N 3], where a new proof of Neumann's contractivity lemma was presented (this proof is included in [KNV IV]).After the work done in [N 3],[Le 1],[Scho] the question of validity of Neumann's contractivity lemma in the plane was settled. As remarked in[KW], the question remained open in spaces of higher dimension. Theorem 3.5 dealing with this question, as well as theorems 3.1, 3.8 are taken from[KN 1]. Consequences of these results for boundary value problems are included in[KN 2]. The least constant q occurring in (10) in § 3 was called the structure constant by Neumann, who gave its estimate for the ellipsoid. For further related investigation cf. [A],[KK], [Schi 1],[Schi 2].

Investigation of the growth of the density of the equilibrium distribution near the vertex of a cube is presented in [Fi].

Convergence of the Neumann series associated with the equation of the form (34) (§3) for general compact operator T in a Banach space is investigated in [Su] (cf. also [Sc] dealing with topological linear spaces).

Convergence of successive approximations in a Green's function approach to the Dirichlet problem for smoothly bounded domains is investigated in [KR].

Theorem 3.13 shows that for every $\mu \in \mathcal{C}_0'(\partial Q)$ there is a harmonic function h on the interior Q of the convex body cl Q such that $N^Q h = \mu$; moreover, this harmonic function can be represented as a potential $U\nu$ with $\nu \in \mathcal{C}'(\partial Q)$. Conditions for representability of harmonic functions on convex bodies by means of potentials are investigated in [Po] where further related references are given.

§ 4 In the historical development the equations

$$(I+\lambda T)f = g \quad (\text{over } \mathcal{C}(\partial G)) ,$$

$$(I+\lambda T)'\mu = \nu \quad (\text{over subspaces of } \mathcal{C}'(\partial G)) ,$$

where $T \equiv T^G$ is the Neumann operator, were first considered for $\lambda = \pm 1$ and a convex body $C = R^m \setminus G$ (compare §3, (37) and (37'), (38) and (38')). Later the investigation extended to complex values of the parameter λ and general smooth domains (cf. [P 1] - [P 3], [HT]) and led to the discovery of Fredholm theorems ([Fr 1] - [Fr 4]) and the development of the so-called Riesz-Schauder theory for functional equations of the second kind with a compact operator T (cf. [RS], [Ri], [R 1], [Scha]). Thus tractability of boundary value problems by the method of integral equations was well established for domains with sufficiently smooth boundaries. Special investigations were needed in case the boundary contained edges and corners (cf. [Ca], [Ar]). Radon introduced a class of curves of bounded rotation ("Kurven beschränkter Drehung") and showed for plane domains bounded by such curves that the Fredholm radius of the corresponding Neumann operator can be estimated in terms of angles enclosed by half-tangents to the boundary at angular points (cf. [R 2]). Radon's investigations were limited to the plane. As remarked in [K 6], Radon's formula for the Fredholm radius of the Neumann operator T^G ceases to hold even for smooth plane domains G submitted to more general assumption $V^G < +\infty$ (cf. (32) in §2). For arbitrary convex body C in R^m $(m \geqq 2)$ the Fredholm radius of T^G $(G = R^m \setminus C)$ was evaluated in [Ne 1] as the reciprocal of $\max\{1 - 2d_C(y);\ y \in \partial C\} =$ $= \max\{d_G(y) - d_C(y);\ y \in \partial C\}$ (cf. 4.1 for notation). The formulae presented in 4.1, 4.2 are taken from [K 2].

For plane domains bounded by curves of bounded rotation, theorem 4.10 guaranteeing uniform continuity of the potential whose generalized normal derivative vanishes goes back to J. Radon, whose method was extented in [BM] to m-space domains satisfying the conditions of 4.10. Somewhat modernized presentation of Radon's method adopted here was used in [Ne 2] in connection with the third boundary value problem.

§ 5 The reader is referred to [Kr], [Pr], [PR], [Ši], [Sch], [We], [Z] and texts on functional analysis concerning equations in linear spaces and the Fredholm alternative.

The quantity $V_r^F(z)$ occurring in 5.3 may be considered as variation in the sense of Banach of the mapping

$$x \longmapsto \frac{x-z}{|x-z|}$$

of $\Omega_r(z) \cap (F \setminus \{z\})$ into Γ. Banach introduced the concept of variation of a continuous transformation in his work [Ba] on rectifiability of curves and surfaces; further generalizations of this concept are studied in [RR].

Finite connectivity of sets $G \subset R^m$ with $V_0^G < \frac{1}{2}$ proved in 5.9 was announced without proof in [K 7]; in [K 2] it is shown that there is a closed $F \subset \partial G$ such that $\mathcal{H}_{m-1}(F) = 0$ and $\partial G \setminus F$ is a locally lipschitzian surface.

Neumann's operator on the space of bounded Baire functions on ∂G and representability of the generalized solution of the Dirichlet problem (with discontinuous boundary data) by means of double layer potentials is investigated in [Ne 3].

Theorem 5.12 describes the range of the operator $N^G\mathcal{U}$ on $\mathcal{C}'(\partial G)$ under the assumption that $G \subset R^m$ satisfies $V_0^G < \frac{1}{2}$. We know that this operator preserves the subspace $\mathcal{C}_c'(\partial G)$ which permits to describe the range of its restriction to this subspace. It is proved in [Ne 4] that $N^G\mathcal{U}$ preserves also the subspace of those $\mu \in \mathcal{C}'(\partial G)$ that are absolutely continuous with respect to the restriction of $\mathcal{H}_{m-1}$ to $\hat{\partial} G$.

In this text we restricted our attention to first two boundary value problems in potential theory and omitted more detailed investigation of the spectrum of the Neumann operator as well as investigation of potentials whose densities are in L_p-classes (cf. [FJL], [CCF]). For application of a variant of the Neumann operator to the third boundary value problem (with a prescribed combination of the Dirichlet and the Neumann boundary condition) the reader is referred to

[Ne 2], [Ne 4], [Ne 5], where further references are given. Similar methods apply to the heat equation (cf. [V 2] for the references) and other equations of mathematical physics (compare [BKM]); we do not enter into the vast related bibliography. We have not considered numerical applications of the method of integral operators and omit the literature on this aspect. As an additional reference we quote the recent book [JS] which appeared when the present text had been completed.

References

[A] L. V. Ahlfors: Remarks on the Neumann-Poincaré integral equation, Pacific J. Math. 3(1952), 271-280.

[Ar] K. Arbenz: Integralgleichungen für einige Randwertprobleme für Gebiete mit Ecken, Promotionsarbeit, Prom. Nr. 2777, Eidgenössische Technische Hochschule Zürich, 1958.

[ASW] G. Anger, B.-W. Schulze, G. Wildenhain: Potentialtheorie (Entwicklung der Mathematik in der DDR, VEB Deutscher Verlag der Wissenschaften Berlin 1974), 428-451.

[BKM] V. M. Babič, M. B. Kapilevič, S. G. Mihlin, G. I. Natanson, P. M. Riz, L. N. Slobodeckij, M. M. Smirnov: Linejnyje uravnenija matematičeskoj fiziki, Moskva 1964.

[Ba] S. Banach: Sur les lignes rectifiables et les surfaces dont l'aire est finie, Fund. Math. 7(1925), 225-237.

[Be] M. Bernkopf: The development of function spaces with particular reference to their origins in integral equation theory, Arch. Hist. Exact Sci. 3(1966/67), 1-96.

[B] M. Brelot: Éléments de la théorie classique du potential, Paris 1961.

[Br 1] J. E. Brothers: A characterization of integral currents, Trans. Amer. Math. Soc. 150(1970), 301-325.

[Br 2] J. E. Brothers: Stokes' theorem, American Journal of Math. 92(1970), 657-670.

[Br 3] J. E. Brothers: Behaviour at the boundary of a solution to Plateau's problem, preprint (Dept. of Math., Indiana Univ., Bloomington).

[BM] Ju. D. Burago, V. G. Maz'ja: Nekotoryje voprosy teorii potenciala i teorii funkcij dlja oblastej s nereguljarnymi granicami, Zapiski naučnyh seminarov LOMI, tom 3, 1967.

[B Me] H. Burkhardt, Fr. Meyer: Potentialtheorie, Encyklopädie der Mathematischen Wissenschaften II A 7b, 464-503, B. G. Teubner, Leipzig 1899-1916.

[CCF] A. P. Calderón, C. P. Calderón, E. Fabes, M. Jodeit, N. M. Riviere: Applications of the Cauchy integral on Lipschitz curves, Bulletin Amer. Math. Soc. 84 (1978), 287-290.

[Ca] T. Carleman: Über das Neumann-Poincarésche Problem für ein Gebiet mit Ecken, Inaugural-Dissertation, Uppsala 1916.

[C] R. Courant: Partial differential equations, New York-London 1962.

[D] I. I. Daniljuk: Nereguljarnyje graničnyje zadači na ploskosti, Moskva 1975.

[DG 1] E. De Giorgi: Nuovi teoremi relativi alle misure (r-1)-dimensionali in uno spazio ad r dimensioni, Ricerche Mat. 4(1955), 95-113.

[DG 2] E. De Giorgi: Su una teoria generale della misura (r-1)-dimensionale in uno spazio ad r dimensioni, Annali di Mat. Pura ed Appl. (4) 36(1954), 191-213.

[Do 1] M. Dont: Non-tangential limits of the double layer potentials, Časopis pro pěst. matematiky 97(1972), 231-258.

[Do 2] M. Dont: On a heat potential, Czechoslovak Math. J. 25(1975), 84-109.

[Do 3] M. Dont: On a boundary value problem for the heat equation, Czechoslovak Math. J. 25(1975), 110-133.

[Do 4] M. Dont: A note on a heat potential and the parabolic variation, Časopis pro pěst. matematiky 101 (1976), 28-44.

[DS] N. Dunford, J. T. Schwartz: Linear operators I, New York-London 1958.

[FJL] E. B. Fabes, M. Jodeit Jr., J. E. Lewis: Double layer potentials for domains with corners and edges, Indiana Univ. Math. J. 26(1977), 95-114.

[Fe 1] H. Federer: The Gauss-Green theorem, Trans. Amer. Math. Soc. 58(1945), 44-76.

[Fe 2] H. Federer: A note on the Gauss-Green theorem, Proc. Amer. Math. Soc. 9(1958), 447-451.

[Fe 3] H. Federer: The (Φ,k) rectifiable subsets of n-space, Trans. Amer. Math. Soc. 62(1947), 114-192.

[Fe 4] H. Federer: Geometric measure theory, Springer--Verlag 1969.

[FF] H. Federer, W. H. Fleming: Normal and integral currents, Ann. of Math. 72(1960), 458-520.

[Fi] G. Fichera: Comportamento asintotico del campo elettrico e della densità elettrica in prossimità dei punti singolari della superficie conduttore, Rendiconti del Seminario Matematico dell' Università e del Politecnico di Torino vol. 32º(1973-74), 111-143.

[Fr 1] I. Fredholm: Sur une nouvelle méthode pour la résolution du problème de Dirichlet, Kong. Vetenskaps.-Akademiens Förh. Stockholm (1900), 39-46.

[Fr 2] I. Fredholm: Sur une classe d'équations fonctionelles, Acta Mathematica 27(1903), 365-390.

[Fr 3] I. Fredholm: Les équations intégrales linéaires, Comptes rendus du Congrés des mathématiciens, Stockholm 1909.

[Fr 4] I. Fredholm: Oeuvres complètes, Litos Reprotryck, Malmö, 1955.

[FK] J. Fuka, J. Král: Analytic capacity and linear measure, Czechoslovak Math. J. 28(1978), 445-461.

[Ga] D. Gaier: Konstruktive Methoden der konformen Abbildung, Springer-Verlag 1964.

[Gou] É. Goursat: Cours d'analyse mathématique, Tome III, 5e ed., Gauthier-Villars, Paris 1956.

[G] N. M. Günter: Die Potentialtheorie und ihre Anwendung auf Grundprobleme der mathematischen Physik, Leipzig 1957 (Übersetzung aus dem Russischen).

[HK] W. K. Hayman, P. B. Kennedy: Subharmonic functions, Academic Press 1976.

[HT] E. Hellinger, O. Toeplitz: Integralgleichungen und Gleichungen mit unendlichvielen Unbekannten,

Encyklopädie der Mathematischen Wissenschaften II C 13, 1335-1597, B. G. Teubner, Leipzig 1923-1927.

[H] L. L. Helms: Introduction to potential theory, Wiley-Interscience, New York 1969.

[JS] M. A. Jaswon, G. T. Symm: Integral equation methods in potential theory and elastostatics, Academic Press, New York 1978.

[J] K. Jörgens: Lineare Integraloperatoren, B. G. Teubner, Stuttgart 1970.

[KK] L. V. Kantorovich and V. I. Krylov: Approximate methods of higher analysis, Interscience, New York 1958 (translated from the Russian).

[Ke] O. D. Kellog: Foundations of potential theory, New York 1929.

[Kle] F. Klein: Vorlesungen über die Entwicklung der Mathematik im 19. Jahrhundert, Springer Verlag 1926.

[KW] R. E. Kleinman, W. L. Wendland: On Neumann's method for the exterior Neumann problem for the Helmholtz equation, Journal of Mathematical Analysis and Applications 57(1977), 170-202.

[Kl] M. Kline: Mathematical thoughts from ancient to modern times, Oxford University Press, New York 1972.

[K I] J. Král: Teorie potenciálu I, Stát. pedagog. nakl. Praha 1965.

[K 1] J. Král: On the logarithmic potential, Comment. Math. Univ. Carolinae 3(1962), $N^{\underline{o}}$1, 3-10.

[K 2] J. Král: The Fredholm method in potential theory, Trans. Amer. Math. Soc. 125(1966), 511-547.

[K 3] J. Král: Graničnoje povedenije potencialov dvojnogo sloja, Trudy Seminara S. L. Soboleva $N^{\underline{o}}$2 (1976), Novosibirsk, 19-34.

[K 4] J. Král: Potentials and boundary value problems, 5. Tagung über Probleme und Methoden der Math. Physik, Wiss. Schriftenreihe der TH Karl-Marx-Stadt 1975, Hft 3, 484-500; Correction of misprints: Comment. Math. Univ. Carolinae 17(1976), 205-206.

[K 5] J. Král: Limits of double layer potentials, Accad.

Nazionale dei Lincei, Rendiconti della Cl. di Sc. fis., matem. e natur., ser. VIII, vol. XLVIII (1970), 39-42.

[K 6] J. Král: On the logarithmic potential of the double distribution, Czechoslovak Math. J. 14(1964), 306-321.

[K 7] J. Král: A note on the Robin problem in potential theory, Comment. Math. Univ. Carolinae 14(1973), 767-771.

[K 8] J. Král: On the Neumann problem in potential theory, Comment. Math. Univ. Carolinae 7(1966), 485-493.

[Kl 1] J. Král, J. Lukeš: On the modified logarithmic potential, Czechoslovak Math. J. 21(1971), 76-98.

[Kl 2] J. Král, J. Lukeš: Integrals of the Cauchy type, Czechoslovak Math. J. 22(1972), 663-682.

[KN 1] J. Král, I. Netuka: Contractivity of C. Neumann's operator in potential theory, Journal of the Mathematical Analysis and its Applications 61(1977), 607-619.

[KN 2] J. Král, I. Netuka: C. Neumann's operator of the arithmetic mean in potential theory, to appear.

[KNV II] J. Král, I. Netuka, J. Veselý: Teorie potenciálu II, Stát. pedagog. nakl. Praha 1972.

[KNV III] J. Král, I. Netuka, J. Veselý: Teorie potenciálu III, Stát. pedagog. nakl. Praha 1976.

[KNV IV] J. Král, I. Netuka, J. Veselý: Teorie potenciálu IV, Stát. pedagog. nakl. Praha 1977.

[KZ] M. A. Krasnosel'skij, P. P. Zabrejko, E. I. Pustyl'nik, P. E. Sobolevskij: Integral'nyje operatory v prostranstvah summirujemyh funkcij, Moskva 1966.

[Kr] S. G. Krejn: Linejnyje uravnenija v banachovom prostranstve, Moskva 1971.

[KR] R. Kress, G. F. Roach: On the convergence of successive approximations for an integral equation in a Green's function approach to the Dirichlet problem, Journal of the Mathematical Analysis and its Applications 55(1976), 102-111.

[La] N. S. Landkof: Foundations of modern potential theory, Springer-Verlag 1972 (translated from the Russian).

[Le 1] H. Lebesgue: Sur la méthode de Carl Neumann, J. Math. Pures Appl. 9e série, XVI (1937), 205-217, 421-423.

[Le 2] H. Lebesgue: En marge du calcul des variations (Une introduction au calcul des variations et aux inégalités géometriques), Monographie de l'Enseignement mathématique N° 12, Institut de Mathématiques, Université de Génève, 1963.

[L] R. Leis: Vorlesungen über partielle Differentialgleichungen zweiter Ordnung, Hochschultaschenbücher, Bibliographisches Institut 165/165a.

[Li] L. Lichtenstein: Neuere Entwicklung der Potentialtheorie. Konforme Abbildung. Encyklopädie der Mathematischen Wissenschaften II C 3, 177-377. B. G. Teubner, Leipzig 1909-1921.

[Lja] A. M. Ljapunov: Raboty po teorii potenciala, Moskva--Leningrad 1949.

[Lon] A. T. Lonseth: Sources and applications of integral equations, SIAM Review 19(1977), 241-278.

[Lo] W. V. Lovitt: Linear integral equations, Mc Graw--Hill Co. New York 1924.

[Lu] J. Lukeš: A note on integrals of the Cauchy type, Comment. Math. Univ. Carolinae 9(1968), 563-570.

[M] J. Mařík: The surface integral, Czechoslovak Math. J. (81) 6(1956), 522-558.

[Ma] E. Martensen: Potentialtheorie, B. G. Teubner, Stuttgart 1968.

[Mi] S. G. Michlin: Integrální rovnice, Přírodovědecké vydavatelství Praha 1952 (translated from the Russian).

[Mi C] C. Miranda: Equazioni alle derivate parziali di tipo ellittico, Springer-Verlag 1955.

[Mi M] M. Miranda: Distribuzioni aventi derivate misure, Insiemi di perimetro localmente finito, Ann. Scuola Norm. Sup. Pisa (3) 18(1964), 27-56.

[Ne 1] I. Netuka: Fredholm radius of a potential theoretic operator for convex sets, Časopis pro pěst. matematiky 100(1975), 374-383.

[Ne 2] I. Netuka: The third boundary value problem in potential theory, Czechoslovak Math. J. 22(1972), 554-580.

[Ne 3] I. Netuka: Double layer potentials and the Dirichlet problem, Czechoslovak Math. J. 24(1974), 59-73.

[Ne 4] I. Netuka: Generalized Robin problem in potential theory, Czechoslovak Math. J. 22(1972), 312-324.

[Ne 5] I. Netuka: An operator connected with the third boundary value problem in potential theory, Czechoslovak Math. J. 22(1972), 462-489.

[N 1] C. Neumann: Untersuchungen über das logarithmische und Newtonsche Potential, Teubner Verlag, Leipzig 1877.

[N 2] C. Neumann: Zur Theorie des logarithmischen und des Newtonschen Potentials, Berichte über die Verhandlungen der Königlich Sachsischen Gesellschaft der Wissenschaften zu Leipzig 22(1870), 49-56, 264-321.

[N 3] C. Neumann: Über die Methode des arithmetischen Mittels, Hirzel, Leipzig, 1887 (erste Abhandlung), 1888 (zweite Abhandlung).

[NE] E. R. Neumann: Studien über die Methoden von C. Neumann und G. Robin zur Lösung der beiden Randwertaufgaben der Potentialtheorie, Teubner Verlag, Leipzig 1905.

[O 1] M. Ohtsuka: Harmonic functions, Lectures at University of Illinois (1966-67) and at Hiroshima University (1971-73).

[O 2] M. Ohtsuka: Modern theory of Newtonian potentials, Hiroshima University 1970-71.

[Pl] J. Plemelj: Potentialtheoretische Untersuchungen, B. G. Teubner, Leipzig 1911.

[dP] N. du Plessis: Introduction to potential theory, Edinburgh 1970.

[P 1] H. Poincaré: Sur les équations de la physique mathématique, Rend. Circolo Mat. Palermo 8(1894), 57-186.

[P 2] H. Poincaré: La méthode de Neumann et le problème de Dirichlet, Acta Mathematica 20(1896), 59-142.

[P 3] H. Poincaré: Théorie du potentiel newtonien, Paris 1899.

[Po] E. Pokorná: Harmonic functions on convex sets and single layer potentials, Časopis pro pěst. matematiky 102(1977), 50-60.

[Pr] S. Prössdorf: Einige Klassen singulärer Gleichungen, Akademie-Verlag, Berlin 1974.

[PR] D. Przeworska-Rolewicz, S. Rolewicz: Equations in Banach spaces, Warszawa 1968.

[RR] T. Rado-P. V. Reichelderfer: Continuous transformations in analysis, Springer Verlag 1955.

[R 1] J. Radon: Über lineare Funktionaltransformationen und Funktionalgleichungen, Sitzber. Akad. Wiss. Wien 128(1919), 1083-1121.

[R 2] J. Radon: Über Randwertaufgaben beim logarithmischen Potential, Sitzber. Akad. Wiss. Wien 128(1919), 1123-1167.

[Ri] F. Riesz: Über lineare Funktionalgleichungen, Acta Math. 41(1917), 71-98.

[RS] F. Riesz, B. Sz.-Nagy: Leçons d'analyse fonctionelle, Budapest 1952.

[Rob] G. Robin: Distribution de l'électricité sur une surface fermée convexe, C. R. Acad. Sci. Paris 104 (1887), 1834-1836.

[Ro] C. A. Rogers: Hausdorff measures, Cambridge University Press 1970.

[Scha] J. Schauder: Über lineare, vollstetige Funktionaloperationen, Studia Math. 2(1930), 183-196.

[Schi 1] M. Schiffer: Problèmes aux limites et fonctions propres de l'équation intégrale de Poincaré et de Fredholm, C. R. Acad. Sci. Paris 245(1957), 18-21.

[Schi 2] M. Schiffer: Fredholm eigenvalues and conformal mapping, Rendiconti di Matematica e delle sue Applicazioni vol. XXII (1963), 445-468.

[Scho] G. Schober: Neumann's lemma, Proc. Amer. Math. Soc. 19(1968), 306-311.

[Sc] J. Schulz: Über die Konvergenz der Neumannschen Reihe in linearen topologischen Räumen, Beiträge zur Analysis 12(1978), 177-183.

[SW] B.-W. Schulze, G. Wildenhain: Methoden der Potentialtheorie für elliptische Differentialgleichungen beliebiger Ordnung, Akademie-Verlag Berlin 1977.

[Sch] Š. Schwabik: Remark on linear equations in Banach space, Časopis pro pěst. mat. 99(1974), 115-122.

[S] L. Schwartz: Théorie des distributions I, II, Actualités Sci. Ind. Nos 1091, 1122, Hermann, Paris.

[Ši] G. E. Šilov: O teoreme Fredgol'ma-Rissa, Vestnik Moskovskogo Univ. 1976, № 1, 59-63.

[So] V. S. Sologub: Razvitie teorii elliptičeskih uravnenij v XVIII i XIX stoletijah, Kiev 1975.

[St] V. A. Steklov : K voprosu o suščestvovanii konečnoj i nepreryvnoj vnutri dannoj oblasti funkcii koordinat, udovletvorjajuščej uravneniju Laplasa, pri zadannyh značenijah normal'noj proizvodnoj na poverhnosti, ograničivajuščej oblast.
Soobščenija Har'kovskogo Matem. Obščestva t. V, Har'kov 1897, 1-32.

[ŠV] J. Štulc, J. Veselý: Connection of cyclic and radial variation of a path with its length and bend (Czech with an English summary), Časopis pro pěst. matematiky 93(1968), 80-116.

[Su] N. Suzuki: On the convergence of Neumann series in Banach space, Math. Ann. 220(1976), 143-146.

[V 1] J. Veselý: On a generalized heat potential, Czechoslovak Math. J. 25(1975), 404-423.

[V 2] J. Veselý: Some remarks on Dirichlet problem, Proceedings of the Summer School "Nonlinear evolution equations and potential theory" held in 1973, Academia, Prague 1975, 125-132.

[We] W. Wendland: Bemerkungen über die Fredholmschen Sätze, Meth. Verf. Math. Phys. 3(1970), 141-176.

[W] J. Wermer: Potential theory, Lecture Notes in Math. vol. 408, Springer-Verlag 1974.

[Y] K. Yosida: Functional analysis, Springer-Verlag 1965.

[You] L. C. Young: A teory of boundary values, Proc. London Math. Soc. (3) 14 A(1965), 300-314.

[Z] A. C. Zaanen: Linear analysis, Amsterdam 1953.

Symbol Index

Symbol Index

Subject Index

Vol. 670: Fonctions de Plusieurs Variables Complexes III, Proceedings, 1977. Edité par F. Norguet. XII, 394 pages. 1978.

Vol. 671: R. T. Smythe and J. C. Wierman, First-Passage Perculation on the Square Lattice. VIII, 196 pages. 1978.

Vol. 672: R. L. Taylor, Stochastic Convergence of Weighted Sums of Random Elements in Linear Spaces. VII, 216 pages. 1978.

Vol. 673: Algebraic Topology, Proceedings 1977. Edited by P. Hoffman, R. Piccinini and D. Sjerve. VI, 278 pages. 1978.

Vol. 674: Z. Fiedorowicz and S. Priddy, Homology of Classical Groups Over Finite Fields and Their Associated Infinite Loop Spaces. VI, 434 pages. 1978.

Vol. 675: J. Galambos and S. Kotz, Characterizations of Probability Distributions. VIII, 169 pages. 1978.

Vol. 676: Differential Geometrical Methods in Mathematical Physics II, Proceedings, 1977. Edited by K. Bleuler, H. R. Petry and A. Reetz. VI, 626 pages. 1978.

Vol. 677: Séminaire Bourbaki, vol. 1976/77, Exposés 489–506. IV, 264 pages. 1978.

Vol. 678: D. Dacunha-Castelle, H. Heyer et B. Roynette. Ecole d'Eté de Probabilités de Saint-Flour. VII-1977. Edité par P. L. Hennequin. IX, 379 pages. 1978.

Vol. 679: Numerical Treatment of Differential Equations in Applications, Proceedings, 1977. Edited by R. Ansorge and W. Törnig. IX, 163 pages. 1978.

Vol. 680: Mathematical Control Theory, Proceedings, 1977. Edited by W. A. Coppel. IX, 257 pages. 1978.

Vol. 681: Séminaire de Théorie du Potentiel Paris, No. 3, Directeurs: M. Brelot, G. Choquet et J. Deny. Rédacteurs: F. Hirsch et G. Mokobodzki. VII, 294 pages. 1978.

Vol. 682: G. D. James, The Representation Theory of the Symmetric Groups. V, 156 pages. 1978.

Vol. 683: Variétés Analytiques Compactes, Proceedings, 1977. Edité par Y. Hervier et A. Hirschowitz. V, 248 pages. 1978.

Vol. 684: E. E. Rosinger, Distributions and Nonlinear Partial Differential Equations. XI, 146 pages. 1978.

Vol. 685: Knot Theory, Proceedings, 1977. Edited by J. C. Hausmann. VII, 311 pages. 1978.

Vol. 686: Combinatorial Mathematics, Proceedings, 1977. Edited by D. A. Holton and J. Seberry. IX, 353 pages. 1978.

Vol. 687: Algebraic Geometry, Proceedings, 1977. Edited by L. D. Olson. V, 244 pages. 1978.

Vol. 688: J. Dydak and J. Segal, Shape Theory. VI, 150 pages. 1978.

Vol. 689: Cabal Seminar 76–77, Proceedings, 1976–77. Edited by A.S. Kechris and Y. N. Moschovakis. V, 282 pages. 1978.

Vol. 690: W. J. J. Rey, Robust Statistical Methods. VI, 128 pages. 1978.

Vol. 691: G. Viennot, Algèbres de Lie Libres et Monoïdes Libres. III, 124 pages. 1978.

Vol. 692: T. Husain and S. M. Khaleelulla, Barrelledness in Topological and Ordered Vector Spaces. IX, 258 pages. 1978.

Vol. 693: Hilbert Space Operators, Proceedings, 1977. Edited by J. M. Bachar Jr. and D. W. Hadwin. VIII, 184 pages. 1978.

Vol. 694: Séminaire Pierre Lelong – Henri Skoda (Analyse) Année 1976/77. VII, 334 pages. 1978.

Vol. 695: Measure Theory Applications to Stochastic Analysis, Proceedings, 1977. Edited by G. Kallianpur and D. Kölzow. XII, 261 pages. 1978.

Vol. 696: P. J. Feinsilver, Special Functions, Probability Semigroups, and Hamiltonian Flows. VI, 112 pages. 1978.

Vol. 697: Topics in Algebra, Proceedings, 1978. Edited by M. F. Newman. XI, 229 pages. 1978.

Vol. 698: E. Grosswald, Bessel Polynomials. XIV, 182 pages. 1978.

Vol. 699: R. E. Greene and H.-H. Wu, Function Theory on Manifolds Which Possess a Pole. III, 215 pages. 1979.

Vol. 700: Module Theory, Proceedings, 1977. Edited by C. Faith and S. Wiegand. X, 239 pages. 1979.

Vol. 701: Functional Analysis Methods in Numerical Analysis, Proceedings, 1977. Edited by M. Zuhair Nashed. VII, 333 pages. 1979.

Vol. 702: Yuri N. Bibikov, Local Theory of Nonlinear Analytic Ordinary Differential Equations. IX, 147 pages. 1979.

Vol. 703: Equadiff IV, Proceedings, 1977. Edited by J. Fábera. XIX, 441 pages. 1979.

Vol. 704: Computing Methods in Applied Sciences and Engineering, 1977, I. Proceedings, 1977. Edited by R. Glowinski and J. L. Lions. VI, 391 pages. 1979.

Vol. 705: O. Forster und K. Knorr, Konstruktion verseller Familien kompakter komplexer Räume. VII, 141 Seiten. 1979.

Vol. 706: Probability Measures on Groups, Proceedings, 1978. Edited by H. Heyer. XIII, 348 pages. 1979.

Vol. 707: R. Zielke, Discontinuous Čebyšev Systems. VI, 111 pages. 1979.

Vol. 708: J. P. Jouanolou, Equations de Pfaff algébriques. V, 255 pages. 1979.

Vol. 709: Probability in Banach Spaces II. Proceedings, 1978. Edited by A. Beck. V, 205 pages. 1979.

Vol. 710: Séminaire Bourbaki vol. 1977/78, Exposés 507–524. IV, 328 pages. 1979.

Vol. 711: Asymptotic Analysis. Edited by F. Verhulst. V, 240 pages. 1979.

Vol. 712: Equations Différentielles et Systèmes de Pfaff dans le Champ Complexe. Edité par R. Gérard et J.-P. Ramis. V, 364 pages. 1979.

Vol. 713: Séminaire de Théorie du Potentiel, Paris No. 4. Edité par F. Hirsch et G. Mokobodzki. VII, 281 pages. 1979.

Vol. 714: J. Jacod, Calcul Stochastique et Problèmes de Martingales. X, 539 pages. 1979.

Vol. 715: Inder Bir S. Passi, Group Rings and Their Augmentation Ideals. VI, 137 pages. 1979.

Vol. 716: M. A. Scheunert, The Theory of Lie Superalgebras. X, 271 pages. 1979.

Vol. 717: Grosser, Bidualräume und Vervollständigungen von Banachmoduln. III, 209 pages. 1979.

Vol. 718: J. Ferrante and C. W. Rackoff, The Computational Complexity of Logical Theories. X, 243 pages. 1979.

Vol. 719: Categorial Topology, Proceedings, 1978. Edited by H. Herrlich and G. Preuß. XII, 420 pages. 1979.

Vol. 720: E. Dubinsky, The Structure of Nuclear Fréchet Spaces. V, 187 pages. 1979.

Vol. 721: Séminaire de Probabilités XIII. Proceedings, Strasbourg, 1977/78. Edité par C. Dellacherie, P. A. Meyer et M. Weil. VII, 647 pages. 1979.

Vol. 722: Topology of Low-Dimensional Manifolds. Proceedings, 1977. Edited by R. Fenn. VI, 154 pages. 1979.

Vol. 723: W. Brandal, Commutative Rings whose Finitely Generated Modules Decompose. II, 116 pages. 1979.

Vol. 724: D. Griffeath, Additive and Cancellative Interacting Particle Systems. V, 108 pages. 1979.

Vol. 725: Algèbres d'Opérateurs. Proceedings, 1978. Edité par P. de la Harpe. VII, 309 pages. 1979.

Vol. 726: Y.-C. Wong, Schwartz Spaces, Nuclear Spaces and Tensor Products. VI, 418 pages. 1979.

Vol. 727: Y. Saito, Spectral Representations for Schrödinger Operators With Long-Range Potentials. V, 149 pages. 1979.

Vol. 728: Non-Commutative Harmonic Analysis. Proceedings, 1978. Edited by J. Carmona and M. Vergne. V, 244 pages. 1979.

Vol. 729: Ergodic Theory. Proceedings, 1978. Edited by M. Denker and K. Jacobs. XII, 209 pages. 1979.

Vol. 730: Functional Differential Equations and Approximation of Fixed Points. Proceedings, 1978. Edited by H.-O. Peitgen and H.-O. Walther. XV, 503 pages. 1979.

Vol. 731: Y. Nakagami and M. Takesaki, Duality for Crossed Products of von Neumann Algebras. IX, 139 pages. 1979.

Vol. 732: Algebraic Geometry. Proceedings, 1978. Edited by K. Lønsted. IV, 658 pages. 1979.

Vol. 733: F. Bloom, Modern Differential Geometric Techniques in the Theory of Continuous Distributions of Dislocations. XII, 206 pages. 1979.

Vol. 734: Ring Theory, Waterloo, 1978. Proceedings, 1978. Edited by D. Handelman and J. Lawrence. XI, 352 pages. 1979.

Vol. 735: B. Aupetit, Propriétés Spectrales des Algèbres de Banach. XII, 192 pages. 1979.

Vol. 736: E. Behrends, M-Structure and the Banach-Stone Theorem. X, 217 pages. 1979.

Vol. 737: Volterra Equations. Proceedings 1978. Edited by S.-O. Londen and O. J. Staffans. VIII, 314 pages. 1979.

Vol. 738: P. E. Conner, Differentiable Periodic Maps. 2nd edition, IV, 181 pages. 1979.

Vol. 739: Analyse Harmonique sur les Groupes de Lie II. Proceedings, 1976–78. Edited by P. Eymard et al. VI, 646 pages. 1979.

Vol. 740: Séminaire d'Algèbre Paul Dubreil. Proceedings, 1977–78. Edited by M.-P. Malliavin. V, 456 pages. 1979.

Vol. 741: Algebraic Topology, Waterloo 1978. Proceedings. Edited by P. Hoffman and V. Snaith. XI, 655 pages. 1979.

Vol. 742: K. Clancey, Seminormal Operators. VII, 125 pages. 1979.

Vol. 743: Romanian-Finnish Seminar on Complex Analysis. Proceedings, 1976. Edited by C. Andreian Cazacu et al. XVI, 713 pages. 1979.

Vol. 744: I. Reiner and K. W. Roggenkamp, Integral Representations. VIII, 275 pages. 1979.

Vol. 745: D. K. Haley, Equational Compactness in Rings. III, 167 pages. 1979.

Vol. 746: P. Hoffman, τ-Rings and Wreath Product Representations. V, 148 pages. 1979.

Vol. 747: Complex Analysis, Joensuu 1978. Proceedings, 1978. Edited by I. Laine, O. Lehto and T. Sorvali. XV, 450 pages. 1979.

Vol. 748: Combinatorial Mathematics VI. Proceedings, 1978. Edited by A. F. Horadam and W. D. Wallis. IX, 206 pages. 1979.

Vol. 749: V. Girault and P.-A. Raviart, Finite Element Approximation of the Navier-Stokes Equations. VII, 200 pages. 1979.

Vol. 750: J. C. Jantzen, Moduln mit einem höchsten Gewicht. III, 195 Seiten. 1979.

Vol. 751: Number Theory, Carbondale 1979. Proceedings. Edited by M. B. Nathanson. V, 342 pages. 1979.

Vol. 752: M. Barr, *-Autonomous Categories. VI, 140 pages. 1979.

Vol. 753: Applications of Sheaves. Proceedings, 1977. Edited by M. Fourman, C. Mulvey and D. Scott. XIV, 779 pages. 1979.

Vol. 754: O. A. Laudal, Formal Moduli of Algebraic Structures. III, 161 pages. 1979.

Vol. 755: Global Analysis. Proceedings, 1978. Edited by M. Grmela and J. E. Marsden. VII, 377 pages. 1979.

Vol. 756: H. O. Cordes, Elliptic Pseudo-Differential Operators – An Abstract Theory. IX, 331 pages. 1979.

Vol. 757: Smoothing Techniques for Curve Estimation. Proceedings, 1979. Edited by Th. Gasser and M. Rosenblatt. V, 245 pages. 1979.

Vol. 758: C. Năstăsescu and F. Van Oystaeyen; Graded and Filtered Rings and Modules. X, 148 pages. 1979.

Vol. 759: R. L. Epstein, Degrees of Unsolvability: Structure and Theory. XIV, 216 pages. 1979.

Vol. 760: H.-O. Georgii, Canonical Gibbs Measures. VIII, 190 pages. 1979.

Vol. 761: K. Johannson, Homotopy Equivalences of 3-Manifolds with Boundaries. 2, 303 pages. 1979.

Vol. 762: D. H. Sattinger, Group Theoretic Methods in Bifurcation Theory. V, 241 pages. 1979.

Vol. 763: Algebraic Topology, Aarhus 1978. Proceedings, 1978. Edited by J. L. Dupont and H. Madsen. VI, 695 pages. 1979.

Vol. 764: B. Srinivasan, Representations of Finite Chevalley Groups. XI, 177 pages. 1979.

Vol. 765: Padé Approximation and its Applications. Proceedings, 1979. Edited by L. Wuytack. VI, 392 pages. 1979.

Vol. 766: T. tom Dieck, Transformation Groups and Representation Theory. VIII, 309 pages. 1979.

Vol. 767: M. Namba, Families of Meromorphic Functions on Compact Riemann Surfaces. XII, 284 pages. 1979.

Vol. 768: R. S. Doran and J. Wichmann, Approximate Identities and Factorization in Banach Modules. X, 305 pages. 1979.

Vol. 769: J. Flum, M. Ziegler, Topological Model Theory. X, 151 pages. 1980.

Vol. 770: Séminaire Bourbaki vol. 1978/79 Exposés 525–542. IV, 341 pages. 1980.

Vol. 771: Approximation Methods for Navier-Stokes Problems. Proceedings, 1979. Edited by R. Rautmann. XVI, 581 pages. 1980.

Vol. 772: J. P. Levine, Algebraic Structure of Knot Modules. XI, 104 pages. 1980.

Vol. 773: Numerical Analysis. Proceedings, 1979. Edited by G. A. Watson. X, 184 pages. 1980.

Vol. 774: R. Azencott, Y. Guivarc'h, R. F. Gundy, Ecole d'Eté de Probabilités de Saint-Flour VIII-1978. Edited by P. L. Hennequin. XIII, 334 pages. 1980.

Vol. 775: Geometric Methods in Mathematical Physics. Proceedings, 1979. Edited by G. Kaiser and J. E. Marsden. VII, 257 pages. 1980.

Vol. 776: B. Gross, Arithmetic on Elliptic Curves with Complex Multiplication. V, 95 pages. 1980.

Vol. 777: Séminaire sur les Singularités des Surfaces. Proceedings, 1976-1977. Edited by M. Demazure, H. Pinkham and B. Teissier. IX, 339 pages. 1980.

Vol. 778: SK_1 von Schiefkörpern. Proceedings, 1976. Edited by P. Draxl and M. Kneser. II, 124 pages. 1980.

Vol. 779: Euclidean Harmonic Analysis. Proceedings, 1979. Edited by J. J. Benedetto. III, 177 pages. 1980.

Vol. 780: L. Schwartz, Semi-Martingales sur des Variétés, et Martingales Conformes sur des Variétés Analytiques Complexes. XV, 132 pages. 1980.

Vol. 781: Harmonic Analysis Iraklion 1978. Proceedings 1978. Edited by N. Petridis, S. K. Pichorides and N. Varopoulos. V, 213 pages. 1980.

Vol. 782: Bifurcation and Nonlinear Eigenvalue Problems. Proceedings, 1978. Edited by C. Bardos, J. M. Lasry and M. Schatzman. VIII, 296 pages. 1980.

Vol. 783: A. Dinghas, Wertverteilung meromorpher Funktionen in ein- und mehrfach zusammenhängenden Gebieten. Edited by R. Nevanlinna and C. Andreian Cazacu. XIII, 145 pages. 1980.

Vol. 784: Séminaire de Probabilités XIV. Proceedings, 1978/79. Edited by J. Azéma and M. Yor. VIII, 546 pages. 1980.

Vol. 785: W. M. Schmidt, Diophantine Approximation. X, 299 pages. 1980.

Vol. 786: I. J. Maddox, Infinite Matrices of Operators. V, 122 pages. 1980.